Pan Study Aids

Human Biology

Donald Reid

Pan Books London and Sydney
in association with **Heinemann Educational Books**

First published 1980 by Pan Books Ltd,
Cavaye Place, London SW10 9PG
in association with Heinemann Educational Books Ltd
ISBN 0 330 26081 2
© Donald Reid 1980
Printed and bound in Great Britain by
Richard Clay (The Chaucer Press) Ltd, Bungay, Suffolk

PAN STUDY AIDS

Titles published in this series

Biology
Book-keeping and Accounts
Chemistry
Commerce
Economics
English Language
Effective Study Skills
French
Geography 1 *Physical and Human*
Geography 2 *British Isles, Western Europe, North America*
History 1 *British*
History 2 *European*
Human Biology
Maths
Physics

Brodies Notes on English Literature

This long established series published in Pan Study Aids now
contains more than 150 titles. Each volume covers one of the major
works of English literature regularly set for examinations.

Contents

Acknowledgements

Thanks are due to Janice Armstrong, Catherine Bett, Sally Bird, Joan Easterbrook, Mary Pocock, Susan Strong, and Sheila Thompson for typing an illegible manuscript with speed and cheerfulness.

Examination questions taken directly from past papers are reprinted by kind permission of the following Examination Boards. The abbreviations used to identify the boards in the text are shown in brackets.

    Associated Examining Board   (AEB)
    Joint Matriculation Board   (JMB)
    Metropolitan Regional Examinations Board   (MREB)
    Oxford Delegacy of Local Examinations   (OX)
    University of Cambridge Local Examinations
       Syndicate   (CAM)
    Welsh Joint Education Committee   (WJ)

Model answers are included for selected examination questions. These are not to be taken as official versions of perfect answers and are not the responsibility of the examination boards concerned. There are many different ways to obtain high marks on a particular question; the answers given in this book are to be taken as examples only.

To the student

The purpose of this book is to help students prepare for GCE O level, AO level, and CSE examinations in human biology and related subjects. The text includes a concise coverage of the main syllabuses for these exams.

The book is not simply a shortened textbook. It is based on a detailed analysis of over one thousand exam questions, together with examiners' reports on papers taken by fifty thousand students. The text is carefully organized under headings which fit the questions most commonly set in the exams.

Attention is drawn to key points, and common errors are described. Advice on how to answer questions, with model answers, is included at the end of each chapter. Practice questions, either taken or adapted from past exam papers, are also incorporated, together with sample mark schemes. These mark schemes are intended as examples of how to write answers which will score maximum marks in the shortest time. However, you should remember that some questions can be approached in different ways and neither the model answers nor the mark schemes should be followed slavishly. The figures in brackets indicate the marks awaded.

When revising, you should pay particular attention to understanding how the body works (physiology) rather than learning the details of its structure. For example, questions on hormonal control of menstruation are much commoner than questions on the structure of the vertebrae. Many graphs are included and explained in the text. These should be studied carefully as they are included in many questions on physiology.

The Exam Boards

The addresses given below are those from which copies of syllabuses and past examination papers may be ordered. The abbreviations (AEB, etc) are those used in this book to identify actual questions.

Associated Examining Board, (AEB)
Wellington House,
Aldershot, Hants GU11 1BQ

University of Cambridge Local Examinations Syndicate, (CAM)
Syndicate Buildings, 17 Harvey Road,
Cambridge CB1 2EU

Joint Matriculation Board,
(Agent) John Sherratt and Son Ltd, (JMB)
78 Park Road,
Altrincham, Cheshire WA14 5QQ

University of London School Examinations Department, (LOND)
66–72 Gower Street,
London WC1E 6EE

Northern Ireland Schools Examination Council (NI)
Examinations Office,
Beechill House,
Beechill Road,
Belfast BT8 4RS

Oxford Delegacy of Local Examinations, (OX)
Ewert Place,
Summrtown,
Oxford OX2 7BZ

Scottish Certificate of Education Examining Board,   (SCE)
(Agent) Robert Gibson and Sons, Ltd,
17 Fitzroy Place,
Glasgow G3 7SF

Southern Universities Joint Board,   (SU)
Cotham Road, Bristol BS6 6DD

Welsh Joint Education Committee,   (WJ)
245 Western Avenue,
Cardiff CF5 2YX

1 Living organisms and their relationships

A basic understanding of this chapter is required by some examination boards. The questions set on this topic require a thorough understanding of the principles involved, but do not require much detailed information. Note that this subject is omitted altogether by some boards.

Characteristics of living organisms

The seven characteristics which distinguish living organisms from non-living organisms are:

1 Respiration All living organisms obtain energy from the breakdown of complex food substances. (**Note:** Respiration is often confused with breathing – p49.)

2 Nutrition All living organisms need a supply of complex molecules as a source of energy and for growth. Animals obtain these by eating other organisms, while green plants manufacture complex molecules from simple molecules.

3 Excretion All living things excrete wastes produced as a result of chemical processes occurring inside their cells. In man, the passing of urine is the chief example of excretion.

4 Growth All living things increase in size at some stage during their life cycle.

5 Reproduction All living things are capable of reproducing themselves at some stage during their life cycle.

6 Sensitivity Living things are able to detect stimuli, i.e. changes in their external or internal environment. For example, animals are sensitive to sound; growing plants are sensitive to light.

7 Response Living things are able to respond to stimuli. Thus animals move on hearing a sudden sound; the shoots of seedlings grow towards light.

Note: Sensitivity and response are sometimes together referred to as **irritability**. Irritability is the ability to respond to a stimulus.

Holophytic and holozoic nutrition

All living organisms require a supply of complex organic molecules, which are then used for growth and respiration inside their cells. Green plants manufacture organic molecules from inorganic molecules, but animals obtain their food by eating other living things.

Holophytic nutrition is the building up of complex organic substances from simple inorganic substances by plants, usually by means of photosynthesis.

Holozoic nutrition is the taking in and digestion of complex organic substances by animals.

Photosynthesis

Photosynthesis is the process by which green plants synthesize (manufacture) glucose from water and carbon dioxide, using the energy from sunlight. It is the ultimate source of energy, organic compounds, and oxygen required for the activities of living things.

Photosynthesis takes place in the presence of the green pigment **chlorophyll** as follows:

1 **Carbon dioxide** from the atmosphere diffuses into the leaf through small pores on the underside of the leaf surface, called **stomata**. It then diffuses into the green cells inside the leaf.
2 **Water** from the soil is taken in by the roots and passed up the stem, through the veins, and into the leaf cells.
3 Inside a green leaf cell there are numerous small circular organelles known as **chloroplasts**, which contain chlorophyll.
4 The **Sun's energy** is trapped by the chlorophyll and used to synthesize glucose from carbon dioxide and water. The following equation summarizes the many steps involved in photosynthesis.

$$\underset{\text{carbon dioxide}}{6\,CO_2} + \underset{\text{water}}{6\,H_2O} + \text{energy from sunlight} \rightarrow \underset{\text{glucose}}{C_6H_{12}O_6} + \underset{\text{oxygen}}{6\,O_2}$$

5 **The glucose** is used by the plant in two ways:
   (a) for respiration, to provide energy in the form of **adenosine triphosphate (ATP)** (p22);
   (b) as a first step in the manufacture of other important substances.

6 **The oxygen** is a waste product and diffuses through the stomata into the atmosphere.

Synthesis of other substances by plants

1 **Sugars,** e.g. sucrose (p77), are made by combining two molecules of glucose. These may be stored in fruits, e.g. apples.
2 **Long chain carbohydrates** (p77), e.g. starch and cellulose, are made by combining many molecules of glucose. These are stored in underground roots (e.g. carrots); in underground stems (e.g. potato tubers); or in seeds (e.g. cereal grains).
3 **Fats or oils** (p78) are made from carbohydrates and are stored in seeds, e.g. nuts.
4 **Amino acids** (p76) are made from glucose, together with nitrates, phosphates, and other salts from the soil. These are taken up by the roots with the water. The amino acids are then combined to form proteins.

Interdependence of plants and animals

(a) Animals are dependent on plants for:
    1 **Energy.** Animals eat plants in order to obtain a supply of complex organic molecules for respiration. By this means, they obtain energy.
    2 **Amino acids.** Animals must also consume plants (or other animals) in order to obtain amino acids for growth, repair, etc.
    3 **Oxygen.** Plants give off oxygen as a by-product of photosynthesis.
    Man is also dependent on plants for wood and natural fibres such as cotton, linen, and flax.
(b) Plants benefit from the presence of animals in the following ways:
    1 **Supply of carbon dioxide.** Animals return carbon dioxide to the atmosphere through respiration (*see* the carbon cycle, p12).
    2 **Supply of inorganic salts,** e.g. nitrates. The dead bodies of animals, and their droppings, provide a rich source of nitrates, etc.
    3 **Dispersal.** Bees disperse pollen and some birds disperse seeds.

Summary of the major differences between plants and animals

1 Plants synthesize food substances by means of photosynthesis. Animals obtain food by eating plants, or animals which feed on plants.

2 Plants usually remain in one place, anchored by their roots. Animals usually seek their food by active movement.

3 Plants give off a gas rich in oxygen by day, and give off a gas rich in carbon dioxide by night. Animals breathe in oxygen and give off carbon dioxide at all times.

(But remember that plants also respire at all times: *see* next section.)

The carbon and nitrogen cycles

The carbon cycle

Carbon circulates continuously through the bodies of plants and animals, as shown in Fig. 1.1. Plants respire continuously just as

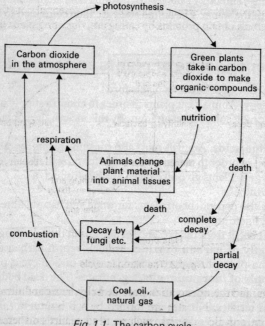

Fig. 1.1 The carbon cycle

animals do. Some of the glucose manufactured during photosynthesis is respired in the plant cells to produce energy, as in animals (p21).

Plant respiration occurs at all times, but is not easy to detect during daylight because the carbon dioxide produced is used up immediately in photosynthesis. However, carbon dioxide can be detected in the gas diffusing out of the leaves at night.

A common mistake made by students is to write, 'Plants breathe out oxygen but animals breathe out carbon dioxide'. In fact, plants give off (they do not breathe) a gas rich in oxygen only in warm, sunny weather, while animals breathe out a mixture of gases containing only 4% carbon dioxide, and as much as 16% oxygen (p55), at all times.

Decay is an important part of the carbon cycle. It is carried out chiefly by microbes, i.e. fungi and bacteria, which break down dead plant and animal remains. Respiration by microbes results in the production of carbon dioxide, as in plant and animal cells.

The nitrogen cycle

Nitrogen is important to living organisms as an essential constituent of proteins. Animals obtain proteins by eating and digesting plant proteins.

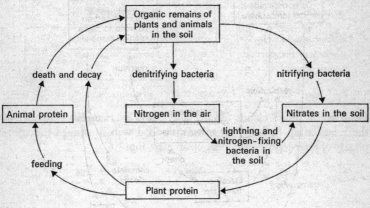

Fig. 1.2 The nitrogen cycle

Plants manufacture amino acids from carbohydrates and nitrogenous salts, such as nitrates. Nitrates are formed by soil **bacteria** (nitrifying bacteria) from organic material, such as dead plants and animals. In addition, **nitrogen-fixing bacteria** found in the soil and in the roots of leguminous plants (e.g. peas and clover) convert nitrogen from the air into amino acids.

Nitrogen also passes back into the atmosphere through the activity of denitrifying bacteria, found especially in waterlogged soil. This production is balanced by the action of lightning which oxidizes atmospheric nitrogen to form nitrogen oxides. These dissolve in rain and are then washed into the soil.

Food chains and food webs

Food chains illustrate the dependence of living organisms on each other for food, e.g.

Plant plankton in the sea → eaten by animal plankton → eaten by herrings → eaten by man.

A **pyramid of numbers** occurs in all food chains with small numbers of animals at the top feeding on larger numbers below them (*see* Fig. 1.3).

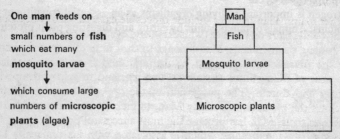

One **man** feeds on
↓
small numbers of **fish**
which eat many
mosquito larvae
↓
which consume large
numbers of **microscopic**
plants (algae)

| Man |
| Fish |
| Mosquito larvae |
| Microscopic plants |

Fig. 1.3 A pyramid food chain

Food webs   Food chains usually provide an oversimplified picture of the relationship between organisms. A food web illustrates the complexity of interdependence more clearly (*see* Fig. 1.4).

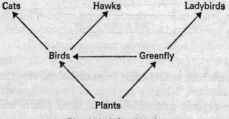

Fig. 1.4 A food web

Man's position among living things

Man has become the dominant species on Earth as a result of his widespread activities, and his ability to alter the environment in his favour.

Our species is the product of at least one thousand million years of evolution, starting with one-celled organisms, followed much later by fish, amphibia, reptiles, and mammals.

Man belongs to the class of vertebrate animals (animals with backbones) known as the **mammals**. The features which together distinguish mammals from other animals are:

1 possession of mammary glands, so that the young can be suckled on milk;
2 the young are born alive;
3 mammals are able to maintain a constant body temperature (homiothermy);
4 possession of fur and sweat glands, both of which contribute to homiothermy.

Other mammalian features include moveable external ears (ear pinnae), long tactile hairs on the face (whiskers), and eyelids to protect the eyes.

Among the mammals, Man belongs to the group of large apes known as the **primates**. These can be distinguished from other mammals because of their ability to walk on two legs and to grip with opposed finger and thumb. The possession of the opposed finger and thumb enables the primates, especially Man, to pick up small objects and use these as tools. Man has become the most successful primate because of his superior intelligence, which is associated with his unusually large and deeply folded cerebral cortex in the brain (p105).

Questions

1 *State four characteristics of living organisms and three major differences between plants and animals.* [p9; 7]
2 *Why are chloroplasts important in the provision of human food?* [p10; 3]
3 *What is the part played by the following in photosynthesis: (a) water, (b) chlorophyll, (c) pores in the surface of a leaf.* [p10; 2 each]
4 *Construct a simple food chain to demonstrate the interdependence of plants and animals. Use four different types of organism as examples* [p14; 4]
5 *Describe carefully with reference to the following sequence:*

Sunlight → wheat → cereal → eaten by → work
*　　　　　plant　　grains　　a human　　output*
*　　　　　leaves　　　　　　being　　　by the*
*　　　　　　　　　　　　　　　　　　human*
*　　　　　　　　　　　　　　　　　　being*

how light energy may be transformed into the kinetic energy of limb movement of that human being. [p10 and p21; 25]   (AEB)

Notes on your answer   Questions of this kind always require careful planning. Start by writing down a plan like this:

1 Photosynthesis
2 Manufacture of starch by the plant
3 Digestion of starch; absorption of glucose
4 Respiration of glucose to produce ATP
5 Contraction of muscle; work done by muscle

6 *Explain in detail how the nitrogen taken in as nitrate from the soil by the root of a broad bean plant may subsequently form part of a man's fingernail.* [p13 and p23; 25]   (AEB)

Answer plan
1 Protein synthesis in plants
2 Digestion of protein and absorption of amino acids
3 Protein synthesis in man

7 (a) *How can primates be distinguished from other mammals?* [p15; 2]
  (b) *Which anatomical feature has encouraged Man's development as a tool maker?* [p15; 1]

2 Cell structure and function

Most living things consist of a highly organized collection of cells. Each cell contains characteristic structures, which can be seen in detail under an electron microscope.

All cells are capable of generating their own supply of energy from the oxidation of carbohydrates (a process known as respiration). Most cells also contain a set of 'coded instructions', known as **genes**, for the manufacture of proteins. The most important proteins in the cells are the enzymes, which act as catalysts for every biological reaction in the body.

Cells, tissues, organs, and systems

The human body is composed of about one hundred million million (10^{14}) individual cells, arranged into various categories known as organs, tissues, and systems.

The cell is the basic unit upon which life is based. Most cells contain a number of structures called **cell organelles** (p19).

Tissues
A collection of cells with similar functions and structure is called a tissue, e.g. nervous tissue (p103).

Organs
A collection of different tissues grouped together in one part of the body to perform the same overall function is called an organ. For example, the lungs contain epithelium, cartilage, red blood cells, and smooth muscles, all combining to carry out the functions of gas exchange (p53).

Systems
A system is a group of organs working together to perform the same overall function. For example, the digestive system contains the mouth, oesophagus, stomach, ileum, colon, and other organs.

Different types of tissue

There are four major kinds of tissue in the body:

1 Epithelial tissues (*see* below).
2 Connective tissue – bone, cartilage, blood, and unspecialized connective tissue (*see* below).
3 Muscle tissue.
4 Nervous tissue.

Unspecialized connective tissue consists of smooth white sheets which bind together different organs. It contains yellow protein fibres for elasticity and white protein fibres for strength.

Epithelial tissues consist of layers of cells joined tightly to each other, and usually separated from other tissues by a thin membrane, known as a **basement membrane**. Epithelia usually lack blood vessels.

Epithelia are found wherever there is a distinct boundary between organs. They form the body's outer and inner linings, e.g. the skin and the linings of glands, tubes, etc. There are two main types of epithelium: **simple epithelium** containing one layer of cells only; and **compound (stratified) epithelium** containing several layers of cells.

There are four main types of simple epithelium (*see* Fig. 2.1):

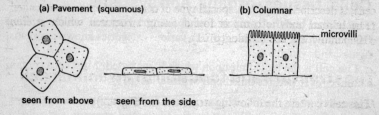

(a) Pavement (squamous)   (b) Columnar — microvilli

seen from above   seen from the side

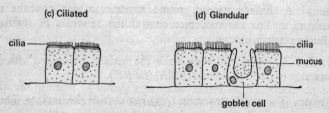

(c) Ciliated   (d) Glandular

cilia — cilia — mucus — goblet cell

Fig. 2.1 Types of simple epithelium

1 **Pavement (squamous) epithelium** consists of thin flat cells. These are found wherever rapid diffusion of substances takes place across a boundary, e.g. the alveoli (p54).

2 **Columnar epithelium** consists of tall column-like cells, found especially in the ileum. The edge of each cell contains numerous folds, known as **microvilli**. These increase the surface area for absorption of digested foods.

3 **Ciliated epithelium** consists of column-shaped cells with numerous minute rhythmically beating hairs, called **cilia**, on their free, outside edge. Cilia are found wherever cells perform the function of moving fluids along a tube. Mucus in the respiratory tubes (p50) and fluid in the oviducts (p127) are moved along by the action of cilia.

4 **Glandular tissue** contains modified epithelial cells, e.g. goblet cells which produce mucus. These are found among the ciliated bronchial epithelium and the columnar intestinal epithelium (p89). **Mucous membrane** is a term used to describe the layer of glandular epithelium, connective tissue, and involuntary muscles which lines the respiratory and digestive systems.

Compound (stratified) epithelium covers areas subject to considerable friction, e.g. the skin, and the linings of the mouth and oesophagus. The epidermis of the skin, which contains three layers of cells, is described on p27. A special type of compound epithelium, called **transitional epithelium**, is found lining structures which stretch considerably, e.g. the bladder (p101).

The structure and function of an animal cell

Most cells contain the following structures (*see* Fig. 2.2):

(a) **Those visible under the light microscope** – magnified up to 1500 times.

Nucleus A densely packed, round structure which contains the chromosomes. The chromosomes control the activities of the cell, especially protein synthesis.

Nucleolus A small round body inside the nucleus, concerned with the manufacture of ribonucleic acid (RNA) (*see* p23).

Centrioles Two small structures lying just outside the nucleus. These are centres for spindle formation during cell division (*see* p149).

Cytoplasm The grey granular jelly which makes up the bulk of the cell, excluding the nucleus.

Cell or plasma membrane The outer boundary of the cell, made of phospholipid and protein molecules. This controls the passage of substances in and out of the cell.

(b) **Those visible under the electron microscope** – magnified up to 500 000 times.

Mitochondria The centres for cell respiration (p21). Here glucose is broken down to produce energy, which is temporarily stored as adenosine triphosphate (ATP), and released as required to the rest of the cell. **The inner membrane** is composed of folds known as **cristae,** which give a larger surface area for the attachment of enzymes concerned with cell respiration. Mitochondria are commonest in active cells, e.g. muscle cells.

Endoplasmic reticulum A complicated network of folded membranes arranged to form tubes and passages. It is concerned with the transport of proteins and other large molecules in and out of the cytoplasm and the nucleus.

Ribosomes Small particles attached to many of the membranes in the endoplasmic reticulum. They are concerned with the manufacture of protein from amino acids.

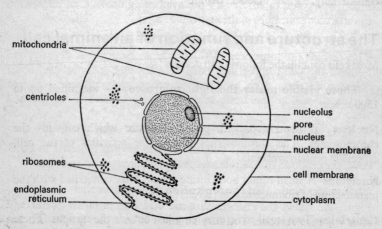

Fig. 2.2 A generalized cell seen under the electron microscope

Nuclear membrane This controls the passage of substances in and out of the nucleus. It has occasional gaps, referred to as pores.

Cell respiration

Cell respiration is the process whereby cells break down sugars such as glucose to produce energy, carbon dioxide, and water. It is also known as **tissue respiration** or **internal respiration**.

The purpose of respiration is to provide energy for the cells to carry out work. Examples of work done in cells include: muscle contraction, protein synthesis (e.g. manufacture of enzymes), and the production of heat. Cells can respire **aerobically** (with oxygen) or **anaerobically** (without oxygen).

Aerobic respiration involves the complete breakdown of carbohydrates such as glucose. This takes place inside the mitochondria in a series of small steps, with release of energy at each stage. Each step is controlled by a separate enzyme. It can be summarized as follows:

$$\underset{\text{glucose}}{C_6H_{12}O_6} \ + \ \underset{\text{oxygen}}{6\,O_2} \rightarrow \underset{\text{water}}{6\,H_2O} \ + \ \underset{\substack{\text{carbon}\\\text{dioxide}}}{6\,CO_2} \ + \ \text{energy}$$

The complete oxidation of one gram molecule of glucose (180 g) produces 2830 kilojoules (kJ) of energy.

Anaerobic respiration involves the partial breakdown of carbohydrates in the absence of oxygen.

During vigorous exercise (e.g. sprinting), muscle cells may not receive sufficient oxygen to oxidize all the available glucose. Cell respiration can still occur but stops at the anaerobic stage:

$$\text{glucose} \rightarrow \text{carbon dioxide} + \text{lactic acid} + \text{energy}$$

This process produces only 118 kJ per gram molecule of glucose, and leads to the build up of poisonous substances, e.g. lactic acid. After vigorous exercise, deep breathing is necessary to provide sufficient oxygen to oxidize the lactic acid completely to carbon dioxide and water.

$$\text{lactic acid} + \text{oxygen} \rightarrow \text{carbon dioxide} + \text{water} + \text{energy}$$

Oxygen debt This is the volume of extra oxygen required to break down the products of anaerobic respiration after vigorous exercise.

Changes in lactic acid concentration during exercise

Figure 2.3 shows the changes in lactic acid concentration during and after vigorous exercise. You may be asked to interpret graphs like this in the exams.

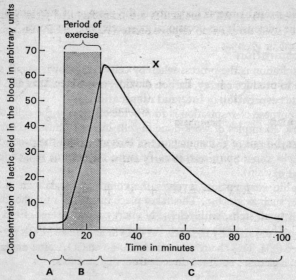

Fig. 2.3 Variations in lactic acid concentration with exercise

During the first 10 minutes (labelled **A**) the subject is at rest so the lactic acid concentration remains low. During the next 10 minutes (labelled **B**) the subject is exercising violently and requires a large quantity of energy rapidly. As she cannot breathe in sufficient oxygen to oxidize the available glucose, her muscle cells break down glucose by anaerobic respiration.

During the final period (labelled **C**) the subject is at rest again and the lactic acid concentration gradually falls, as it is oxidized to carbon dioxide and water. Rapid breathing during this period provides the oxygen needed to 'pay back' the oxygen debt incurred during exercise.

At the point on the graph marked **X** the lactic acid concentration is still rising although exercise has ceased. This occurs because lactic acid is still passing out of the muscles into the blood.

Energy transfer

The energy produced by respiration is used to manufacture a chemical called **adenosine triphosphate** (**ATP**). This consists of the substance adenosine joined to three phosphate groups:

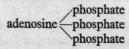

During respiration, ATP molecules are manufactured from a simpler chemical called **adenosine diphosphate (ADP)**. ADP contains only two phosphate groups:

$$\text{adenosine}\!\!\begin{array}{c} P \\ \diagdown \\ \diagup \\ P \end{array} \; + \; P \; + \underset{\substack{\text{(from} \\ \text{respiration)}}}{\text{energy}} \longrightarrow \text{adenosine}\!\!\begin{array}{c} P \\ \diagdown \\ -P \\ \diagup \\ P \end{array}$$

$$\qquad\quad \text{ADP} \qquad\qquad \text{phosphate} \qquad\qquad\qquad\qquad \text{ATP}$$

ATP diffuses out of the mitochondria into all parts of the cell. When energy is required by the cell to carry out work, ATP is rapidly broken down again.

$$\text{ATP} \longrightarrow \text{ADP} \; + \; P \; + \; \text{energy}$$

How cells manufacture proteins

Proteins are required by the body for growth and repair. Each protein is composed of a long chain made from amino acids. Amino acids (p76) are small molecules obtained directly from digested food.

DNA and RNA

Role of DNA Each cell contains a set of stored (base) codes for the making of proteins. The code is made up of a chemical called **deoxyribosenucleic acid (DNA)**, found inside the chromosomes.

Role of RNA A different nucleic acid, called **ribonucleic acid (RNA)**, is found chiefly in the ribosomes in the cytoplasm. RNA molecules are able to fit together with strands of DNA in the correct order specified by the DNA base code. The code is then carried out to the ribosomes in the following stages:

1 Separate pieces of RNA become lined up alongside a DNA molecule inside the nucleus, and thus form a complete molecule of **messenger RNA**. This carries the code for a complete protein.

2 The messenger RNA leaves the nucleus through the pores in the nuclear membrane.

3 Messenger RNA becomes attached to a ribosome in the cytoplasm.

4 In the cytoplasm, small molecules of a different type of RNA, known as **transfer RNA** become linked each to a single amino acid.

5 Transfer RNA molecules become lined up along the messenger RNA molecule in the correct sequence.

6 The amino acids become linked together by peptide bonds to form polypeptides (p76), and, finally, complete proteins. The formation of

a peptide bond requires expenditure of energy; this is supplied by ATP, formed during cell respiration.

Relationship between chromosomes, genes, and DNA

Each cell contains a set of chromosomes (p149). The chromosomes carry genes (p24), which confer particular characteristics (e.g. brown or blue eye colour) on the individual.

Genes are composed essentially of DNA. Each gene contains the code for the manufacture of a particular protein, in the form of long strands of DNA.

For example, the gene for blue eye colour is composed of slightly different DNA from the gene for brown eye colour. So each gene contains the code for the manufacture of a slightly different protein.

Importance of DNA

1 Inherited characteristics are passed on from one generation to the next in the form of DNA, which is carried on the chromosomes in the sperm and ova.
2 DNA is essential for the synthesis of all cell proteins because it contains the codes needed for their manufacture.

Enzymes – functions and characteristics

Enzymes are proteins which act as organic catalysts. Enzymes are capable of accelerating the rate at which reactions take place, but are not themselves used up.

Without enzymes, chemical reactions inside the body would occur very slowly. For example, the breakdown of starch to maltose sugar (p87) inside the mouth normally takes days; if the enzyme salivary amylase is present, it only takes a few minutes.

Properties of enzymes

Specificity Each type of enzyme catalyses only one kind of reaction. Thus amylases will break down starch but not proteins.

Optimum temperature Enzymes work best at particular temperatures. Human enzymes work best between 35 °C and 40 °C, i.e. blood temperature. Below 35 °C, they work more slowly, but speed up if warmed. For every 10 degree rise in temperature, the rate of reaction is doubled.

Above 45 °C, rates of reaction become slower again. Most enzymes become inactive at about 60 °C. Heating an enzyme above 60 °C causes a permanent change in the shape of the molecule. The enzyme is then said to be **denatured**.

(**Note** Enzymes are *not* living things and so cannot be 'killed' by heat. This is a common mistake made in exam answers.)

Optimum pH Enzymes are also sensitive to changes in acidity and alkalinity. For example, salivary amylase in the mouth functions best at pH 7·5 (slightly alkaline) but pepsin functions best in the stomach at pH 2·0 (acid).

Intracellular enzymes The great majority of enzymes are intracellular, i.e. they function inside cells. These include the enzymes which catalyse cell respiration inside the mitochondria.

Extracellular enzymes These function outside cells: for example, digestive enzymes (p90) are made inside cells, but must be released into the gut before they become active.

Questions

1 *Explain the meaning of the following terms and give an example of each:*
  (a) cell organelle, (b) tissue, (c) organ, (d) organ system [p17; 8]
2 *How can epithelia be distinguished from other kinds of tissue?* [p18; 2]
3 *In each of these cases, state:*
   (i) the type of epithelium present;
   (ii) how the structure of that epithelium is related to its function. [p19; 3 each]
   (a) ileum, (b) alveoli, (c) bronchi and trachea, (d) capillaries,
   (e) epidermis, (f) bladder.
4 *Describe the main structures of an animal cell.* [p19; 20]
5 *Explain the importance of tissue respiration to the body.* [p21; 4]
6 *Briefly state the functions and relationships to each other of adenosine triphosphate (ATP) and adenosine diphosphate (ADP) in tissue respiration.* [p23; 5]   (AEB)
7 *How does a human muscle cell obtain energy from the sugar and oxygen supplied in the blood?* [p21; 9]

Answer guide:
Sugar and oxygen diffuse [1] out of the blood into the muscle cells. In the mitochondria [1], they combine together, in the presence of enzymes [1], to produce energy as shown:

$$6\ O_2 + C_6H_{12}O_6 \longrightarrow 6\ CO_2 + 6\ H_2O + energy$$

oxygen   glucose   carbon   water
                       dioxide

[1 for stating the correct chemical formula and 1 for stating the equation correctly in words.]

This process is called cell respiration [1]. The energy derived from it is used to manufacture ATP [1]:

$$\text{energy} + \text{ADP} + \text{P} \longrightarrow \text{ATP} \quad [\text{1 for correct formula.}]$$

The ATP produced in this way provides the energy for muscle contraction [1].

8 *Write short notes on DNA, genes and chromosomes.* [p23; 3 each]
   (CAM)
9 *Where are nucleic acids found in a cell? Briefly explain their importance in inheritance.* [p23; 6]   (AEB)
10 (a) *Name three important properties of enzymes.* [p24; 3]
   (b) *Why are enzymes less efficient above 45 °C?* [p24; 1]

3 The skin and temperature regulation

The skin: structure and functions

The skin is an organ which covers the entire surface of the body. It is composed of an outer **epidermis**, an inner **dermis**, and a layer of **subcutaneous fat** below the dermis. (Fig. 3.1)

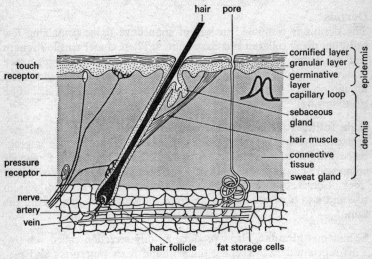

Fig. 3.1 Section through the skin

Epidermis
Cornified layer This consists of flat, dead cells full of granules composed of the protein **keratin**. It forms a tough, almost waterproof layer, which prevents the entry of microbes and reduces loss of water

from the body. The cells are constantly worn away and replaced from the granular layer beneath.

Granular layer This consists of living cells produced by the germinative layer beneath. These cells are continuously converted to cornified cells. Keratin is deposited inside them, and they lose their nuclei and become flattened in shape.

Germinative (Malpighian) layer This is an active layer of cells which constantly divides to produce new epidermis. Sweat glands, sebaceous glands, and hair follicles are produced from infoldings of the epidermis, which reach deep into the dermis. The germinative layer contains the dark pigment **melanin,** which absorbs ultra violet radiation.

The epidermis is an example of compound (or stratified) epithelium (p18) because it contains several layers of cells. It varies in thickness, being especially thick and cornified on the soles of the feet to resist pressure, and relatively thin on the sensitive lips.

Dermis

The dermis is composed mainly of connective tissue containing few cells, but many collagen and elastic fibres. These confer tensile strength and elasticity. White blood cells (p58) called **macrophages** move slowly through the dermis where they engulf microbes and other foreign objects. Various structures are found in the dermis, including the following:

Hair follicles These are deep pits composed of granular and germinative layers. The germinative cells multiply and become impregnated with keratin, so building up a single hair. Each hair is a cylinder composed of dead cells; the hair grows as new cells are added at the 'root'. Contraction of the **hair erector muscles** causes the hairs to rise and also squeezes out sebum from the sebaceous glands onto the skin.

Sebaceous glands These produce an oily secretion called **sebum**. This repels water, so keeping the skin and hairs waterproof and preventing loss of water from the body. Sebum also helps to prevent the multiplication of microbes on the skin.

Sweat glands Sweat glands consist of coiled tubes, lined by cells which absorb fluid from the surrounding tissues and capillaries. This fluid (sweat) is then passed through the coiled tubes onto the surface of the skin. Sweat is 99% water, with 0·3% salt, especially sodium chloride, and minute amounts of urea and lactic acid.

In hot weather, a man performing heavy work may lose up to 10 litres of sweat per day and 30 g of sodium chloride. It is not sufficient merely to drink water to replace the lost fluid: salt tablets must also be taken. Otherwise the salt and water balance of the body may be upset, leading to painful cramp in the muscles.

The capillaries in the dermis supply food and oxygen and remove waste substances. The arterioles leading to these capillaries play a major role in temperature regulation.

Sense organs The dermis of non-hairy skin contains various types of nerve ending, capable of responding to touch, heat, cold, and pressure. These nerve endings, called **receptors** (p114), provide information about the nature of our surroundings, e.g. whether a surface is rough or smooth, hot or cold. The receptors also have a **protective function**: if strongly stimulated, they produce the sensation of pain.

Subcutaneous fat
This is found beneath the dermis. It acts both as a food reserve and an insulating layer, to prevent loss of heat.

Summary of functions of the skin

1 It protects the body against dehydration, invading microbes, mechanical damage, and damage due to ultra violet rays and poisonous chemicals.
2 It contains receptors sensitive to heat, cold, touch, and pressure. These assist in the protective function.
3 It plays a major role in temperature control (p30).
4 It has a minor role as an excretory organ (p96). Minute amounts of urea and lactic acid are excreted. (Water and sodium chloride are also lost, but cannot be regarded as excreted products.)

Temperature regulation

Humans, like all mammals and birds, are able to maintain a **constant body temperature**. This permits normal activity in all weathers. Animals which cannot maintain their body temperature, e.g. reptiles and insects, lose heat and become sluggish and inactive during cold weather. Constant body temperature therefore permits man to be active in a wide range of climates from the Arctic to the tropics.

Heat production

Heat is produced by chemical reactions within the cells, especially cell respiration (p21) in the liver and muscles, and is then distributed around the body by the circulatory system. **The subcutaneous fat** acts as an insulator and so reduces loss of heat.

Heat loss

Heat is lost from the body to the atmosphere by radiation, convection, and conduction (p192). These are the normal methods by which any warm object loses heat to its surroundings: most heat is lost from the body in this way. Cooling also occurs by loss of latent heat caused by the evaporation of sweat. As sweat evaporates, latent heat is removed from the body. The output of sweat depends upon the body temperature.

The means by which heat is *lost* should not be confused with the *control* of body temperature. Heat production and heat loss are normally kept in balance, so producing a constant body temperature. However, if the temperature of the blood begins to rise or fall, various mechanisms operate to control heat loss or gain.

Control of body temperature
Overheating

This may be caused by hot weather, vigorous exercise, a high fever, or exposure to excessive solar radiation (e.g. sunbathing).

If the temperature of the blood rises, temperature receptors (p106) in the hypothalamus in the brain send nervous impulses to the skin. These impulses cause the following changes:

1 Vasodilation The arterioles in the skin dilate (swell), so increasing the flow of blood through the skin. This leads to increased loss of heat through the epidermis by convection and radiation. (**Note:** The arterioles in the skin do not move up and down: they merely dilate or constrict.)

2 Sweating The sweat glands are stimulated by nervous impulses to secrete additional sweat.

Overcooling

This may be caused by exposure to cold weather, especially in strong winds, because moving air carries heat away from the body rapidly by convection. When a fall in blood temperature is detected by the hypothalamus, the following changes occur:

1 Vasoconstriction, i.e. narrowing of the skin arterioles. This reduces the flow of blood to the skin and so minimizes heat loss.

2 Shivering This is a spasmodic reflex contraction of the voluntary muscles. As a result, increased energy production occurs in the muscles, and body temperature rises.

3 Increased metabolic rate (p82). The hypothalamus stimulates the thyroid gland to produce the hormone thyroxine (p111). This increases the metabolic rate, resulting in greater production of heat.

4 Sweating is reduced This minimizes the loss of latent heat.

5 Contraction of hair erector muscles In hairy mammals and birds, the fur or feathers are raised from the skin. This traps a thicker layer of warm air between the skin and the fur. Since still air conducts heat badly, a layer of fluffed out fur or feathers is an effective insulator.

In man, the hairs are mostly too short to produce this insulating effect. The contraction of the erector muscles merely causes the swellings known as 'goose pimples'.

Hypothermia is the medical name for a marked drop in body temperature, due to exposure to cold. The symptoms include a fall in body temperature to 33 °C or lower, feelings of drowsiness, and a slow pulse. If the victim falls asleep, the symptoms become worse, with the circulation slowing down, the blood pressure falling, and the temperature dropping still further until death occurs.

Hypothermia is especially found during a severe winter among **old people** living alone who either lack money for fuel, or are incapable of arranging adequate heating. In addition, the temperature regulation mechanism is less effective in old age.

Hypothermia may also occur in **very young babies** whose temperature regulating mechanism has not yet fully developed. A baby left in a cold room for long periods gradually loses heat, and becomes still, with swollen hands and feet. Immediate hospital treatment is required if this occurs. To prevent hypothermia in the very young, their rooms should be kept warm at all times.

Young people and adults occasionally suffer from hypothermia as a result of exposure to extreme cold when out walking or climbing. **If stranded in an exposed place** (e.g. a mountainside at night), the following steps should be taken to prevent hypothermia:

1 Make a shelter of some kind, to gain protection from the wind.
  Moving air rapidly carries away body heat by convection.
2 Keep the muscles active to produce body heat.

3 Keep awake, because metabolic rate falls with sleep.
4 Remove wet clothing. As damp clothes dry, they cool the body by absorbing latent heat in the same way as sweating.
5 Put on layers of dry clothes or blankets, if available. Many layers are preferable to one or two thick garments.
6 Do not take alcohol. It creates a temporary feeling of warmth by causing dilation of the arterioles in the skin. This leads to further loss of heat and so hastens death.

Experiments with temperature control

The body's main 'thermostat' is situated in the hypothalamus and is sensitive to changes in blood temperature but not to changes in the outside air temperature. If a person is placed in a hot room at 45 °C, and then drinks a large quantity of iced water, this has surprising effects on the body, as shown in Fig. 3.2. This experiment is often referred to in exam questions.

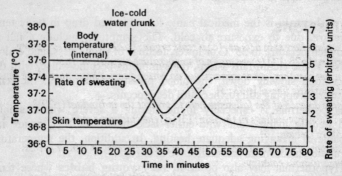

Fig. 3.2 The effects of drinking ice-cold water

1 **Internal body temperature falls.** This is due to the cooling effect of the cold water on the blood around the stomach; as a result, the temperature of the whole circulatory system falls slightly.
2 **The rate of sweating falls.** When the cooled blood reaches the hypothalamus, sweating is reduced to prevent further cooling.
3 **Skin temperature rises.** If sweating is reduced, the skin is no longer kept cool. Since the outside temperature is so hot (45 °C), skin temperature must rise.

After a few minutes, the iced water gains heat from the body and all three measurements return to normal.

Clothing and temperature regulation

To conserve heat, extra layers of clothes should be worn. Warm air is trapped between the layers, so providing the same insulating effect as hair or fur. Many layers of thin clothes are more effective than one or two layers of thick garments. The open texture of string vests results in the trapping of a particularly large volume of air, so they are especially useful in cold climates.

Choice of clothing in hot weather:

1 **Cotton** clothes should be worn in preference to garments made with **synthetic** fibres. Cotton and other natural fibres allow water from perspiration to pass through them; but synthetics prevent water loss and so restrict loss of heat through sweating.
2 **Light coloured** garments should be chosen because they reflect heat. Dark colours absorb heat.
3 **Loose fitting** clothes allow circulation of air around the body, and so speed up loss of heat by sweating.

Questions

1 *Explain how skin acts as: (a) a sense organ;* [p29; 3] *(b) an excretory organ;* [p28;] 4] *(c) an organ of temperature regulation.* [p30; 6]
   In hot weather, what advantages do cotton fabrics have over those made of synthetic fibres? [p33; 3]   (WJ)
2 *How do each of the following help to protect the individual* (p29)?
   (i) the nerve endings in the skin; [3] *(ii) the epidermis of the skin;* [3] *(iii) sebum.* [3]
3 *Suppose you were preparing to go on a long journey in winter across mountainous country.*
   (a) What precautions would you take against hypothermia, before setting off? [p31; 3]
   (b) State two symptoms of hypothermia. [2]
   (c) What are the effects of hypothermia on the blood and circulation? [3]
4 *Why are light-coloured, loose-fitting clothes most suitable for wear in hot, sunny weather?* [p33; 2]
5 *Why is it hard to keep cool in a hot, damp climate?* [4] See below.

Answer guide

The body's cooling mechanism depends chiefly on the evaporation of sweat [1] which absorbs latent heat from the skin [1] as it evaporates. In a humid atmosphere, the air is too damp to absorb further moisture [1], so that the sweat does not evaporate [1].
(**Note** the importance of finding four separate facts to earn all four marks.

4 The skeleton, muscles, and movement

The amount of detail required in this topic varies considerably from syllabus to syllabus, e.g. the structure of the vertebrae is often omitted. Check your syllabus carefully.

The skeleton: functions and structure

General functions

1 The skeleton **supports** the body, allowing an upright posture.
2 The skeleton **protects** vital organs, e.g. the heart, lungs, and brain.
3 The skeleton is essential for **movement** because the bones provide a firm surface for the attachment of muscles. The bones also act as levers.
4 The cells of the red marrow inside the bones **manufacture red blood corpuscles** and white blood cells (p58).

(**Hint** In exams, it is important to give clear answers. It is not sufficient to write 'protection, support, movement' as three functions of the skeleton. In each case, an example must be given, e.g. 'The skeleton aids movement, because the bones provide a firm surface for muscle attachment.')

Axial and appendicular skeleton

The skull, backbone, ribs, and breastbone (sternum) form a central axis, known as the **axial skeleton.** The bones of the limbs, with the pelvic and shoulder girdles, form the **appendicular skeleton** (*see* Fig. 4.1.)

The skull consists of twenty-two bones, all fused together except for the mandible (lower jaw). There is a large hole underneath the skull (called the **foramen magnum**) for the entry of the spinal cord. The functions of the skull are mainly protective, as is shown in Fig. 4.2.

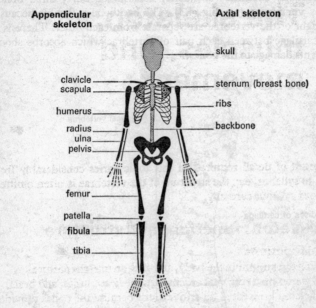

Appendicular skeleton

- clavicle
- scapula
- humerus
- radius
- ulna
- pelvis
- femur
- patella
- fibula
- tibia

Axial skeleton

- skull
- sternum (breast bone)
- ribs
- backbone

Fig. 4.1 The human skeleton

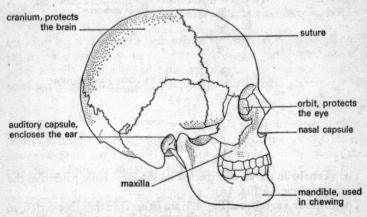

- cranium, protects the brain
- suture
- orbit, protects the eye
- nasal capsule
- auditory capsule, encloses the ear
- maxilla
- mandible, used in chewing

Fig. 4.2 The skull

The vertebral column (backbone) consists of thirty-three bones known as **vertebrae** (*see* Fig. 4.3). The backbone is flexible because twenty-eight of the vertebrae are separate from each other. There is a **disc** of cartilage between each pair of vertebrae which absorbs shock and prevents damage to the bones.

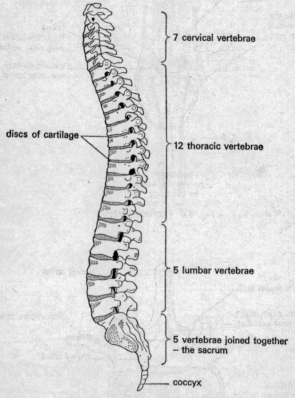

discs of cartilage

7 cervical vertebrae

12 thoracic vertebrae

5 lumbar vertebrae

5 vertebrae joined together
– the sacrum

coccyx

Fig. 4.3 The backbone

The **verbebrae** are all constructed on the same basic pattern as the thoracic vertebrae (*see* Fig. 4.4).

The **lumbar vertebrae** (Fig. 4.5) are larger than the other vertebrae because they support the most weight. The **cervical** (**neck**) **vertebrae** (Fig. 4.6) are the smallest, since they only support the head. The five

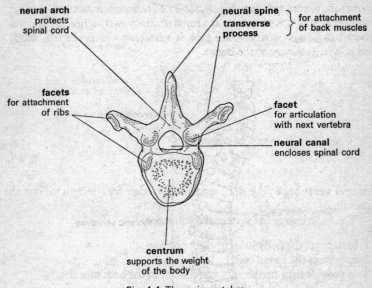

Fig. 4.4 Thoracic vertebra

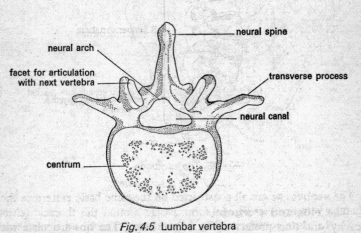

Fig. 4.5 Lumbar vertebra

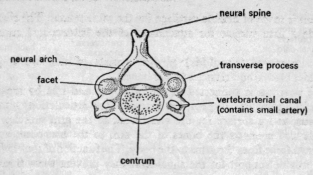

Fig. 4.6 Cervical vertebra

sacral vertebrae (Fig. 4.7.) are fused together for strength to form the sacrum.

The functions of the vertebral column are:

1 to support the body in an upright position;
2 to protect the spinal cord;
3 to provide firm surfaces for attachment of the back muscles.

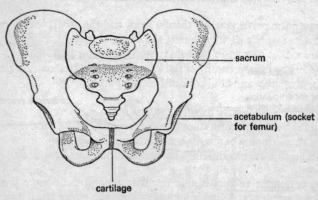

Fig. 4.7 The pelvic girdle

The ribs and sternum form a cage around the thoracic (chest) cavity, and thus protect the heart and lungs. The ribs articulate with the thoracic vertebrae; each can move to a small extent to allow for the movements made during breathing. The junction between the ribs and sternum is made of flexible cartilage for the same reason. The ribs also provide a firm surface for attachment of the **intercostal muscles** (p50).

The **pectoral (shoulder) girdle** consists of the two scapulas (shoulder blades) and clavicles (collar bones).

The **scapula** is connected to the backbone and ribs by means of powerful ligaments. It provides a socket for articulation with the humerus, and a surface for the attachment of the arm muscles. The scapula also connects the bones of the arm to the backbone, and so transmits force from the arms to the body. The function of the **clavicles** is to provide support for the shoulders: they prevent them from collapsing inwards.

The **pelvic girdle (pelvis)** forms a ring of bone, fused to the sacrum for strength (Fig. 4.7). It consists of two halves, a right and left pelvis, each composed of three fused bones. Its functions are:

1 to provide a socket for articulation with the femur;
2 to transmit the thrust from the leg bones to the backbone, and to transmit the weight of the body from the backbone to the leg bones;
3 to provide a surface for muscle attachment;
4 to protect the organs of the lower abdomen.

In women the pelvis is wider than in men, and the cartilage connection at the front opens slightly during the birth of a child. Both of these features aid the passage of the baby's head during birth (p138).

The limbs consist mainly of the long bones, i.e. the femur, humerus, radius, ulna, tibia and fibula. Their functions are:

1 to provide a surface for muscle attachment;
2 to act as levers – they produce a large movement of the hand or foot for a small amount of muscle contraction;
3 to transmit thrust to and from the hands and feet, and the body.

In the **arms** the **radius** is able to rotate around the ulna, e.g. when using a screwdriver. Gliding joints between the many small bones in the hand allow for flexibility in the fingers (see Fig. 4.8). The position of the thumb, opposite the other fingers, allows small objects to be picked up and branches to be grasped.

The bones in the **legs** are larger and stronger than those in the arms, as they support the weight of the body. The **patella** (knee cap) protects the knee joint and also improves the leverage of the muscles which

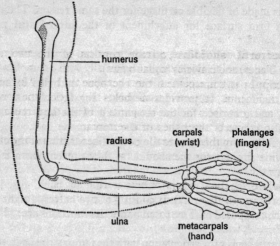

Fig. 4.8 Bones of the arm

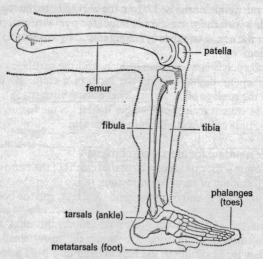

Fig. 4.9 Bones of the leg

straighten the knee (*see* Fig. 4.9). The bones of the foot allow for movement at the ankle, but are mainly designed for strength, to support the weight of the body.

Joints

Types of joint

(a) **At fixed or fibrous joints,** no movement can occur between the bones. The bones fit closely together, usually with a complicated interlocking pattern, e.g. the sutures between the bones in the cranium. Fibrous connective tissue holds the edges of the bones in place. Other examples include the joints between the sacrum and the pelvis, and those between the sacral vertebrae.

(b) **At slightly movable or cartilaginous joints,** some movement is possible: e.g. between the centra of two adjacent vertebrae or between the two pubic bones at the front of the pelvis. A disc made of fibrocartilage (p45) is present between the bones.

(c) **At freely movable or synovial joints,** greater freedom of movement is possible. These joints contain an inner **synovial membrane,** which secretes **synovial fluid** to lubricate the joint. Fibrocartilage is present between the bones to absorb shock and to smooth the movement between the bones.

   The entire joint is enclosed by a **capsule** of fibrous tissue, continuous with the outer layer of bone (the **periosteum,** p45). The bones are held together by strong elastic **ligaments** to prevent dislocation.

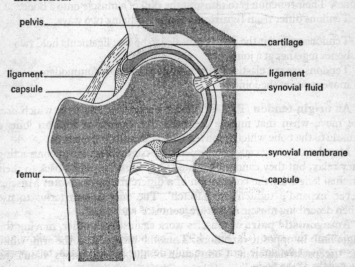

Fig. 4.10 Section through a synovial joint

There are four kinds of **synovial joint**, each permitting movement in a different way:

(i) **Gliding joints** allow one bone to glide over the next, e.g. the bones of the wrist and ankle.

(ii) **Hinge joints** allow movement in one plane only, e.g. the elbow and knee joints.

(iii) **Ball and socket joints** allow movement in any plane, because they have a rounded head fitting into a cup-shaped socket. Examples include the joints between the femur and pelvis, and between the humerus and scapula.

(iv) **Pivot joints** allow the movement of one bone around another. The first cervical vertebra (the atlas) pivots around the odontoid process on the second cervical vertebra (the axis). This allows the head to be turned from side to side, e.g. when shaking the head.

Movement

The skeleton is moved by the contraction of muscles. Muscles are joined to bones by means of tendons.

Tendons are extensions of the connective tissue which surrounds muscles. They are closely attached to the periosteum covering the bones. Their function is to transmit the pull of a muscle onto a bone.

Tendons differ from **ligaments** in the following two ways.

1 Tendons transmit the pull of muscles to bones: ligaments hold two bones together at a joint.

2 Tendons are not elastic: ligaments are elastic to accommodate movements at the joint.

An **origin tendon** (Fig. 4.11) joins a muscle to the bone which *does not* move when that muscle contracts. An **insertion tendon** joins a muscle to the bone which *is* moved when the muscle contracts.

Muscles can only cause movements by contracting. After contracting they relax, but they cannot elongate. They must be pulled back to their original length by the contraction of a different muscle. **Note:** Muscles never 'expand', 'tighten', or 'stretch'. The only correct terms to use when describing muscle action are 'contract' and 'relax'.

Antagonistic pairs of muscles work against each other, moving the same limb in opposite directions. Thus the biceps bend the arm, while the triceps straighten it. Contraction of the triceps brings about the elongation of the biceps.

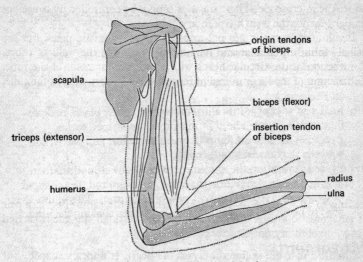

Fig. 4.11 Antagonistic muscles of the forearm

The **flexor** muscle in an antagonistic pair (e.g. the biceps) bends the limb, while the **extensor** muscle (e.g. the triceps) straightens it.

Types of muscle

1 Voluntary, or skeletal muscle The muscles which move the bones are composed of bundles of fibres surrounded by a sheath of connective tissue. As they are under the direct control of the conscious parts of

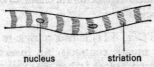

Fig. 4.12 Voluntary muscle fibre

the brain, they are known as voluntary muscles. Since their fibres appear striped under the microscope, they are also called striated or striped muscles. Within the fibres there are no obvious cell membranes; each fibre contains many nuclei.

2 Involuntary or smooth muscle is found in the walls of tubular structures such as the arteries, arterioles, the intestines, and the uterus. It is composed of sheets of separate cells, which do not appear striated under the microscope. These muscles cannot be controlled by conscious thought (hence the term 'involuntary').

Involuntary muscles are not arranged in antagonistic pairs, but are usually found in circular and longitudinal sheets in the walls of tubes. Contraction of the circular fibres will reduce the diameter of the tube. Contraction of the longitudinal fibres produces a shortening of the tube,

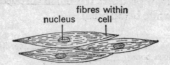

Fig. 4.13 Involuntary muscle

leading to an increase in diameter. This gives the same effect as a pair of antagonistic muscles. Movement of food through the gut is carried out by involuntary muscles (*see* peristalsis, p90).

3 Cardiac muscle is found only in the heart. It appears striped, but there are cross connections between the fibres not found in striated

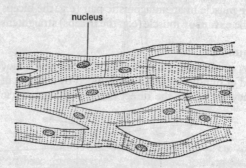

Fig. 4.14 Cardiac muscle

muscles. Heartbeat is an involuntary action and cannot be speeded up by conscious thought. Cardiac muscle can maintain a steady beat for the whole of a lifetime without becoming fatigued.

The structure of bone

A typical long bone (Fig. 4.15) ends in a layer of **cartilage**; this smooths the movement at the joints and prevents damage to the bone.

The outermost layer of bone is a tough fibrous sheath called the **periosteum**. This contains blood vessels which supply food and oxygen to the cells inside the bone. The shaft is composed of **compact bone**, reinforced with **cancellous (spongy) bone** at each end.

Red marrow is found in the spaces within the cancellous bone, and **yellow marrow**, which consists mostly of fat, is found inside the bone cavity. The red marrow produces red blood corpuscles and also certain types of leucocyte (white blood cells).

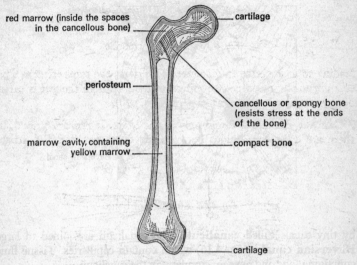

Fig. 4.15 Structure of a long bone

Microscopic structure of cartilage and bone
Cartilage (or gristle) consists of a matrix composed of a gelatinous material secreted by cartilage cells known as **chondrocytes**. The matrix may also contain non-elastic collagen fibres for strength and elastic fibres for flexibility.

The three main types of cartilage are:

1 **Hyaline cartilage,** a clear smooth cartilage found in the joints, and also in the trachea.
2 **Fibrocartilage,** a tough white cartilage containing collagen fibres for strength. This is found especially in the discs between the vertebrae and in the tendons.

3 **Elastic cartilage,** which contains elastic fibres for flexibility, e.g.
  the ear pinna, and nasal cartilage.

Bone is a harder material than cartilage because it consists of a
matrix containing a heavy deposit of calcium phosphate. Collagen
fibres are also present and confer tensile strength.

Bone is a living tissue which contains cells called **osteocytes**. These
lie in small spaces in the bone; the spaces are connected to each other

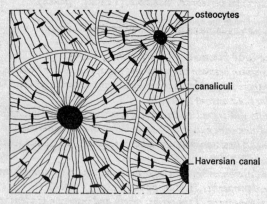

Fig. 4.16 Structure of bone

by tiny canals called **canaliculi.** The canaliculi are joined to larger
Haversian canals (Fig. 4.16) which contain capillaries. Tissue fluid,
containing food and oxygen, circulates in the canaliculi.

The development of bone (ossification)
In the foetus, the long bones are at first made largely from cartilage. The
conversion of cartilage to bone is called **ossification**. It begins during
development in the womb and finishes up to 20 years later when bone
growth finally ceases.

Care of the feet
Since the bones of young children contain a high proportion of cartilage,
they can be distorted easily by tight clothing or small shoes. Babies
should be allowed to go barefoot as much as possible, and children's
shoes must always allow for growth. At all ages, narrow shoes may
cause painful patches of hard skin called **corns.** The wearing of high-
heeled shoes forces the toes forward and may cause permanent dis-

tortions, such as hammer toes and painful outgrowths of bone at the base of the big toe known as **bunions.**

Questions

1 *Given a diagram of the skeleton, label the following:*
sternum; sacral vertebrae; femur; fibula; humerus; ulna; meta-tarsals;
carpals; cervical vertebrae; mandible. [1 each]   (MREB)

(Note: Some syllabuses do not require these names.)

2 (i) *State* **four** *functions of the skeleton.* [p34; 4]
  (ii) *Distinguish between the axial and the appendicular skeleton.* [p35; 2]
3 *Why are there several kinds of synovial joint in the body? Give* **two**
  reasons for your answer. [p42; 2]
4 *Give a fully illustrated description of a synovial and a cartilaginous*
  joint. and quote one example of each. What are the differences in both
  structure and function, between a hinge joint and a ball and socket
  joint? [p41; 25]
5 *State the differences between ligaments and tendons, with reference to*
  (i) physical properties, (ii) functions. [p42; 4]   (CAM)
6 *Explain the meaning of the terms: origin and insertion tendons; flexor*
  and extensor muscles. [p42; 4]
7 *Describe the actions involved in (i) raising the hand to touch the*
  shoulder; (ii) lowering the hand again. [p43; 25] See below.

Answer guide
This full length question requires a detailed answer which should include these points:

1 A description of the action of the flexor (the biceps muscle), including mention of the tendons and the lever action of the forearm bones, in raising the forearm. To be illustrated by Fig. 4.11.
2 Mention of the role of the hinge joint at the elbow, and its structure (cartilage, synovial fluid). Illustrate by Fig. 4.10 adapted in shape to show the ulna and the humerus.
3 To touch the shoulder, the fingers must be bent, so the action of the tendons in the hand, controlled by flexor muscles in the lower arm, should be mentioned.
4 Return to the original position is achieved by using the antagonists for each of the muscles mentioned above. The extensor muscles in the lower arm straighten the fingers, while the extensor in the upper arm

(triceps) straightens the arm. If in doubt, practise the movement on yourself. You can feel the different muscles contracting.

Remember   Human biology is one of the few subjects where you can bring your own study aid with you – yourself.

5 Breathing and gas exchange

Breathing and respiration There is much confusion over the use of these terms. **Breathing,** sometimes called external respiration, is the act of forcing air in and out of the lungs by the use of the diaphragm and intercostal muscles. **Respiration** is the chemical process occurring inside cells whereby food substances are oxidized to produce energy (p21). It is also called tissue respiration, cell respiration or internal respiration.

Structure of the respiratory system

In the **nose,** air passes through narrow passages supported by cartilage. These passages increase the surface area of the epithelium (*see* below) in contact with the air. As a result, incoming air is warmed and moistened by close contact with the cells. **Olfactory cells,** sensitive to smell (p123), are located on the upper surface of the nasal cavity. They are stimulated by the presence of chemicals in the air.

Ciliated epithelium (p19) lines the inside of the nose and the respiratory passages generally. The beating of the cilia directs dust particles out of the nostrils. The epithelium also produces **mucus** in which dust and microbes are trapped. Mucus from the respiratory passages is propelled by the cilia to the throat where it is swallowed.

The trachea (windpipe) is the tube carrying air from the nose to the lungs. It contains incomplete rings of cartilage which keep it permanently open.

The **epiglottis** is a flap of cartilage which prevents food entering the trachea during swallowing (p87). The **larynx,** or voice box, contains two folds of tissue called the **vocal cords.** These can be stretched by the vocal muscles so that they vibrate and produce sounds when air passes over them. The cords are thicker in men than in women producing a deeper note.

The trachea divides into two **bronchi**, which branch into numerous tiny **bronchioles**, terminating in thin-walled, lobed sacs called **alveoli** in which gas exchange takes place.

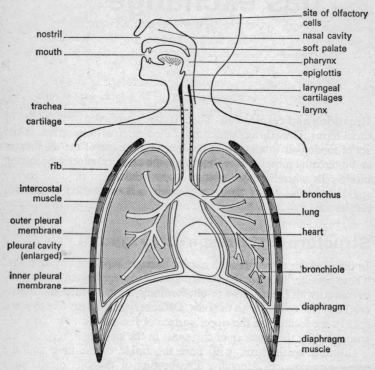

Fig. 5.1 Sagittal section of head and thorax

The mechanism of breathing

Inspiration (inhalation or breathing in) is brought about by the action of voluntary muscles surrounding the thorax (chest). There are no voluntary muscles inside the lungs themselves.

During inspiration
1 The muscles in the outer part of the diaphragm contract, pulling the diaphragm downwards.
2 The intercostal muscles contract to raise the ribs.

3 Both these movements increase the volume of the lungs and so reduce the pressure of the air inside them. As a result, air at atmospheric pressure is able to enter the lungs.

During expiration

1 The diaphragm and intercostal muscles relax, allowing the abdominal organs to push the diaphragm upwards. The lungs contract because of their elasticity, and this is sufficient to expel air during a gentle breath.
2 Forced expiration is brought about by the contraction of the muscles in the body wall surrounding the abdomen. This pushes the stomach against the diaphragm, so exerting pressure on the thoracic cavity and forcing air out of the lungs.

(Note: The **diaphragm** is essentially a large circular sheet of fibrous tissue, with muscles situated around the circumference only. It is **not** simply a large sheet of muscle.)

The lungs will not expand during inspiration unless the surrounding chest wall is airtight. For this reason, the lungs are enclosed in a double membrane, called the pleural membranes.

The **pleural membranes** line the entire surface of the lungs and the inside of the thoracic cavity. This renders the cavity airtight. If the membranes are punctured on one side, one lung will collapse and remain useless until the wound is healed. The membranes contain a thin fluid which lubricates the lung surface, thus allowing free movement of the lungs.

A model to demonstrate breathing

A **bell jar model** of the chest (Fig. 5.2) is sometimes used to demonstrate breathing. When the rubber sheet is pulled down, the volume of the space inside the jar ('thoracic cavity') is increased. This reduces the pressure on the balloons ('lungs') and allows air to enter the balloons, which then inflate slightly. This model has the following defects:

1 The balloons are too small in proportion to the size of the model. In real life, the lungs fill the thoracic cavity.
2 The space inside the thoracic cavity is therefore too large.
3 The glass sides of the jar cannot move so that rib movement cannot be shown.
4 In life, the diaphragm curves upwards into the thoracic cavity even when it is contracted. In the model it is often pulled down to demonstrate 'inspiration' – this does not normally occur.

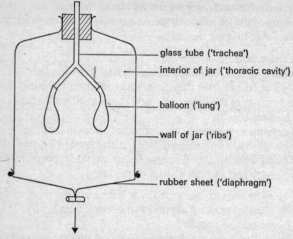

Fig. 5.2 Bell jar model of the chest

Lung Capacity

The total capacity of the lungs is about 5·5 litres in an adult man, but even the deepest breath cannot exceed 4·5 litres. Various terms are used to describe the volumes of air taken in or expelled, etc. (*see* Table 5·1).

A **spirometer** can be used to record changes in volume at inspiration or expiration. This may produce traces on a revolving drum as in Fig. 5·3. You may be asked to interpret a similar figure in the examination.

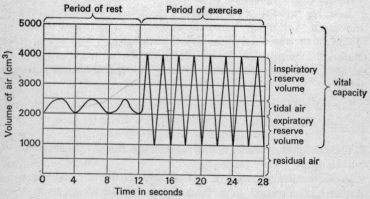

Fig. 5.3 Chest movements during breathing

Table 5.1 Lung capacity in a typical adult

Name	Volume in cubic centimetres (cm^3)	Definition
1 **Tidal air**	500	The volume of air breathed in or out during quiet breathing.
2 **Inspiratory reserve volume** or complementary air	1600	The extra volume of air breathed in during a deep breath, in addition to the tidal air.
3 **Expiratory reserve volume** or supplementary air	1600	The extra volume of air breathed out during a deep breath, in addition to the tidal air.
4 **Dead space**	170	The volume of air which fills the trachea and bronchi and is expelled without reaching the alveoli.
5 **Vital capacity**	3870	The maximum amount of air which can be expelled by forcible expiration following deep inspiration (the sum of 1–4 above).
6 **Residual air**	1600	The volume of air which remains in the respiratory system even after the deepest expiration.
7 **Total lung capacity**	5470	The maximum volume of air present in the lungs (the sum of 5–6 above).

Note:

1 The volumes given vary greatly in different individuals (*see* Fig. 5.3) and you may find different figures quoted in examination papers.

2 The terms used above are only needed for AO and some O level exam papers.

Gas exchange

During the short period when air is inside the alveoli, dissolved oxygen diffuses from the air through the wall of the alveoli into the blood, and dissolved carbon dioxide diffuses from the blood into the air. However this exchange is incomplete so that even exhaled air contains ample oxygen for respiration (*see* Table 5·2).

Table 5.2 Approximate composition of inspired and expired air

	Inspired (i.e. atmospheric) Air	Expired air
Oxygen	21%	16%
Carbon dioxide	0.03%	4%
Nitrogen	79%	79%
Water vapour	varies	always saturated
Other differences	cooler; may contain dust	warmer; most dust particles removed

The **alveoli** (Fig. 5·4) are adapted for rapid diffusion, like the villi in the ileum, in the following ways:

1 They have a moist surface in which gas molecules dissolve.
2 They are thin-walled to permit rapid passage of molecules.
3 Their rounded shape greatly increases the surface area for absorption.
4 Each receives a rich blood supply.

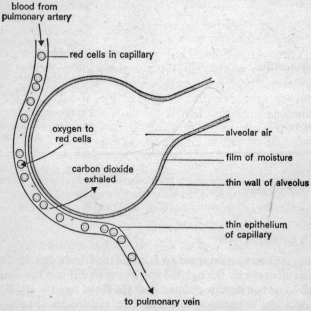

Fig. 5.4 Gas exchange in an alveolus

Oxygen diffuses from the alveolar air, through the film of moisture and the alveolar epithelium, into the blood capillary. This occurs because the concentration of oxygen in alveolar air is higher than the concentration in the blood, i.e. there is a diffusion gradient from the alveolus to the capillary. The oxygen then combines with the haemo-globin in the red blood cells (p58) and is carried through the capillaries into the pulmonary veins to the heart.

Carbon dioxide diffuses from the plasma into the alveolus because there is a diffusion gradient in the opposite direction. Deep breathing allows air to remain in the alveoli for a longer period. This increases the volume of gases exchanged and so increases the volume of carbon dioxide in the expired air.

The air inside the alveoli themselves is not fully exchanged with each breath. Alveolar air therefore contains more carbon dioxide and less oxygen than expired air:

	Alveolar air	*Expired air*
Carbon dioxide	6%	4%
Oxygen	14%	16%

Control of breathing rate

A steady rate of breathing is maintained at about 20 breaths per minute by reflex action, under the control of a group of cells called the **respiratory centre** in the medulla of the brain (p107).

The cells in the respiratory centre are sensitive to increases in the carbon dioxide concentration in the blood. They respond by stimu-lating deeper and faster breathing. Deeper breathing increases the concentration of carbon dioxide in the expired air. Eventually this causes the carbon dioxide concentration in the blood to fall and the respiratory centre cells then cease to stimulate increased breathing. This is an example of **homeostasis** (p60).

Effects of air pollution

The intake of dust particles (e.g. from smoke) over a long period damages the delicate lining of the respiratory tubes. Smoke particles slow down the action of the cilia, so that mucus is not removed. As a result, microbes and dust particles accumulate, increasing the risk of infections and of damage to the tissues. Certain types of dust and pollen may also trigger off an allergic reaction; e.g. hay fever.

Medical conditions associated with the breathing in of smoke, dust, etc., include the following:

1 **Bronchitis** (inflammation of the bronchi and trachea). This is especially common in heavily industrialized urban areas. The symptoms include increased mucus production, leading to a partial blockage of the breathing tubes, which makes breathing difficult. The passing of the Clean Air Act has reduced the incidence of bronchitis in industrial towns.
2 **Respiratory infections,** e.g. coughs, colds, and pneumonia, are all higher among those who live in smoky areas.
3 **Pneumoconiosis** is the name given to a group of diseases caused by breathing in various forms of dust (e.g. silicosis). These conditions are often associated with a particular occupation, e.g. quarrying or mining. Silica particles (as in silicosis), asbestos fibres, and coal dust all cause damage to the lungs, and stimulate increased production of mucus. This leads to coughs, breathlessness, bronchitis, and an increase in respiratory infections. Asbestos fibres may also cause lung cancer, even up to 20 years after exposure.

The wearing of masks, ventilation of factories, and damping down of dusty surfaces help to prevent these conditions.

Cigarette smoking is particularly dangerous because:

1 The smoke particles slow down the action of the cilia; in heavy smokers, the cilia may even disappear altogether. Smokers therefore cough frequently in order to bring up mucus. They are also likely to contract **bronchitis** and other respiratory infections. Repeated damage to the lungs from infection and coughing may lead to loss of alveoli, so that the victim becomes short of breath. This condition is called **emphysema.**
2 Certain chemicals in tobacco smoke are carcinogenic (cancer promoting), and so may cause **lung cancer.** This may take many years to develop.
3 Carbon monoxide in tobacco smoke combines with haemoglobin to form carboxyhaemoglobin (p183). This causes the deposition of fats on the walls of blood vessels, and leads to an increased risk of **heart attacks**.
4 **Smoking during pregnancy** may lower the birth weight of the baby and therefore leads to the death of the infant in some cases.
5 Smoking prevents the healing of gastric and duodenal **ulcers.**

However, giving up smoking gradually restores health (except when permanent lung damage has occurred). Ten years after giving up, most ex-smokers are as healthy as non-smokers.

Other forms of pollution are discussed on p182.

Questions

1 *What is the difference between respiration and breathing?* [p49; 2]
2 (a) *How are the internal structure of the nose and nasal cavities adapted to their functions?* [p49; 10)
   (b) *Why is it preferable to breathe through the nose instead of the mouth?* [3]
3 (a) *Draw large, fully labelled diagrams to show* [p50]:
        (i)   *the structure and position of those parts of the body, including the lungs, concerned with breathing;* [10]
        (ii) *the structures concerned with the exchange of gases.* [5]   (AEB)
4 *Describe the mechanism by which air is:*
   (a) *inspired;* [p50; 5]
   (b) *expired.* [4]
5 *Make a list of the differences between inspired and expired air.* [p54; 8]
6 *What is the cause of pneumoconiosis?* [p56; 1]
   Give **two** *causes of this disease* [2] *and state what precautions can be taken to prevent pneumoconiosis.* [3]
7 *Explain the reasons why smoking is dangerous to health.* [p56; 10]

6   The circulatory system

Composition and functions of blood

Composition of blood (Fig. 6.1)

Blood is mainly composed of a fluid called plasma, which contains red cells, white cells, and platelets in suspension. The volume of blood in the average adult is about 4 litres.

Red cells (**erythrocytes**) are manufactured in the red marrow and survive about 4 months before destruction in the liver. Their chief function is to carry oxygen from the lungs to the tissues. Their structure is suited to this function because the biconcave shape provides a large surface area for the absorption of oxygen, and the cytoplasm contains the protein haemoglobin, which has a powerful affinity for oxygen.

Haemoglobin combines readily with oxygen to form oxyhaemoglobin, especially in areas of high oxygen concentration, such as the lungs. Oxyhaemoglobin is unstable and breaks down easily in areas of low oxygen concentration, e.g. within the tissues. The breakdown of oxyhaemoglobin is also favoured by the acid conditions found in active tissues. This acidity is caused by the presence of carbon dioxide produced during respiration.

Anaemia is an illness caused by a shortage of red cells in the blood. This may be due to lack of iron or of vitamin B_{12} in the diet (p79). It may also result from loss of blood, e.g. as a result of heavy menstrual periods.

White cells (**leucocytes**) differ from red cells as follows:

1 White cells are larger, with diameters varying between 8 and 15 μm, compared to 7·5 μm for red cells.
2 White cells are fewer in number, with only 8000 per cubic millimetre of blood, compared to $5\frac{1}{2}$ million red cells per cubic millimetre.

(a) Red cells

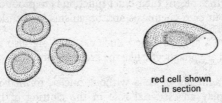

red cell shown
in section

(b) White cells

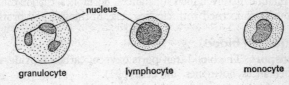

granulocyte   lymphocyte   monocyte

(c) Platelets

Fig. 6.1 Blood cells

3 White cells possess a nucleus, red cells do not.
4 White cells vary in shape; all healthy red cells are biconcave discs.

There are three major types of white cell: granulocytes (70% of all white cells), lymphocytes (23%); and monocytes (7%).

Granulocytes (also called neutrophils or polymorphs) are made in the red bone marrow. They have an irregular shape with granular cytoplasm and a nucleus divided into two or three lobes. Their function is to ingest microbes and dead cells, a process known as **phagocytosis**. Granulocytes are capable of independent movement and can pass through minute gaps in capillary walls to enter the tissues. They accumulate around areas of injury and infection, where they accelerate healing by ingesting invading microbes and damaged tissues.

Lymphocytes are smaller than granulocytes and have a round nucleus which occupies most of the cell. They are manufactured in the lymph nodes (p73) and the spleen; their chief function is to produce antibodies (p61) to destroy foreign chemicals and organisms in the blood.

Monocytes have a slightly lobed nucleus; they are also made in the lymph nodes and are phagocytic like the granulocytes.

Platelets are fragments of cells without nuclei, formed in the bone marrow. They play an essential role in the clotting of blood (p61).

Plasma is the liquid component of blood. It is a pale yellow fluid, composed mainly of water (90%) and salt (3·5%). Since a major function of plasma is the transport of substances around the body, many different chemicals are present in plasma, either dissolved or in suspension. These include dissolved food substances, hormones, carbon dioxide, and the proteins concerned with clotting, e.g. fibrinogen and prothrombin.

Functions of blood

1 Transport The blood transports oxygen, carbon dioxide, dissolved food substances, hormones, and urea (p94).

2 Distribution of heat Heat is produced in the body as a result of respiration by active tissues, chiefly the liver and muscles. It is distributed around the body by the blood, so maintaining an even temperature throughout.

3 Temperature control The opening and closing of arterioles in the skin plays a major role in the maintenance of an even body temperature (p30).

4 Maintenance of a constant internal environment Most human cells can function only within a narrow range of conditions. A relatively constant temperature and pH, together with a steady supply of food and oxygen, and the removal of wastes, are essential for optimum functioning of the cells. The maintenance of steady conditions is called **homeostasis**; the circulation of the blood plays an important role in homeostasis by creating constant conditions throughout the body.

5 Healing and prevention of infection When a blood vessel is damaged, the platelets and the protein fibrin combine to form a clot. This seals the wound, preventing loss of blood and entry of microbes. Any microbes which do enter the body are engulfed by phagocytic leucocytes assisted by the action of antibodies.

The mechanism of clotting

1 When a blood vessel is damaged and exposed to air, platelets in the bloodstream adhere to the damaged areas and form a **platelet plug**.

2 The platelets also release a chemical which causes the involuntary muscles in the walls of the damaged blood vessels to constrict, so reducing the flow of blood.

3 If the damage is too great to be sealed by a plug, the platelets release an enzyme called **thrombokinase** (or thromboplastin).

4 **Thrombokinase** converts an inactive blood protein called **prothrombin** into an active enzyme called **thrombin**.

5 Thrombin converts **fibrinogen** into **fibrin**, in the presence of calcium salts (ions) in the plasma. Fibrin forms a clot, composed of a mesh of fibrin with trapped blood cells inside it.

6 Finally, the clot shrinks and hardens to form a **fibrous scab** which protects the tissues beneath, while healing takes place.

Immediately after a clot has formed, a clear liquid may be seen escaping from it. This consists of plasma lacking fibrinogen or fibrin, and is known as **serum**. The clotting of blood is a complex process involving many stages. If clotting was a simpler process it might occur accidentally inside undamaged vessels, so blocking them and leading to death.

When blood is collected from blood donors, it is liable to clot because it is exposed to the air. To prevent this, **sodium citrate** solution is added. This removes calcium ions from solution, precipitating them as solid calcium citrate, and so inhibits the action of thrombin.

Prevention of infection

If microbes enter the body through a cut, phagocytic white cells migrate out of the capillaries and engulf the invading microbes. This leads to the formation of **pus**, a mixture of dead white cells and damaged tissues. In addition, antibodies are manufactured by the lymphocytes.

Antibodies are substances produced by lymphocytes and other cells to destroy 'foreign' chemicals, foreign tissues, or invading microbes in the body. **Antigens** are foreign microbes, chemicals, or tissues which stimulate the production of antibodies. Each antibody is specific to a particular antigen. Thus, antibodies against one type of flu virus will not attack a different kind of flue virus. Different kinds of antibody have different effects.

1 Some antibodies destroy microbes by dissolving their cell membranes.
2 Antibodies called **agglutinins** cause microbes to clump together, so limiting the extent of their invasion and facilitating their destruction by granulocytes.
3 Antibodies called **anti-toxins** are manufactured in order to neutralize the toxins (poisons) produced by microbes.

There is often a rise in the body's temperature following infection. This is caused by a general increase in metabolic rate, as the body increases its production of white cells and antibodies. During illness, temperature rises in the early evening and falls at night.

Immunity

Following recovery from an invasion by microbes, antibodies may remain in the blood for months or longer and will give protection (immunity) from any further invasion by the same antigen. If immunity is acquired following an infection, it is referred to as **natural immunity**. If it is acquired by vaccination, it is called **artificial immunity**.

Vaccines are made from dead or weakened microbes, or from an inactive form of their toxin. Injection of these weakened antigens causes the production of antibodies against them: further doses containing stronger versions of the antigen can then be injected to build up greater immunity. A booster injection may be needed periodically to maintain the level of protection.

A schedule of recommended vaccinations is set out in Table 6.1.

Table 6.1 Timing of recommended vaccinations

Age	Vaccine
3 months	Initial dose of whooping cough, diphtheria, and tetanus vaccines combined as one triple vaccine. Initial dose of polio vaccine.
6 months	Second dose of triple vaccine and polio vaccine.
11 months	Third dose of triple vaccine and polio vaccine.
13 months	Measles vaccination.
5 years	Booster doses for diphtheria, tetanus, and polio.
12–13 years	BCG vaccine for tuberculosis if required. Rubella vaccination for all girls.
15–19 years	Boosters for polio and tetanus.

(**Note** The words 'immunization', 'vaccination', 'inoculation', and 'injection' are all used to describe this process, but immunization or vaccination are the most appropriate terms.)

Vaccination was first discovered by **Edward Jenner** in 1796 when he injected cowpox virus, a harmless virus similar to smallpox, into a young boy. He later injected the boy with smallpox; when the youth failed to catch the disease it became evident that he was now immune to smallpox.

Active immunity is the type of artificial immunity gained from vaccination in which the body produces its own antibodies. This may give protection for many months or years. **Passive immunity** is artificial immunity obtained by direct injection of antibodies, usually obtained from the serum of a horse. Passive immunity confers only temporary protection.

Blood groups

Blood can be classified into different groups according to the proteins present on the red cells. The most important blood groups are those known as **A, B, AB,** and **O**.

People with blood group A carry an antigen known as A on their red cells; group B carry an antigen called B, people with group AB carry both A and B, while people with group O carry neither. In addition, each group possesses naturally occurring antibodies in the plasma which cause the red cells of other groups to agglutinate (clump together). These antibodies are therefore referred to as **agglutinins**, while the antigens are known as **agglutinogens**.

When **blood transfusions** are needed following loss of blood, it is essential to know the donor's and the recipient's blood groups. If

Table 6.2 Summary of compatibility of blood groups

Blood group	Agglutinogen (antigen) present on red cells	Agglutinin (antibody) present in plasma	Can donate blood to	Can receive blood from
A	A	anti-B (β)	A and AB	A and O
B	B	anti-A (α)	B and AB	B and O
AB	A and B	none	AB only	all groups
O	none	anti-A and anti-B (α and β)	all groups	O only

blood containing a 'foreign' agglutinogen is given, the recipient's blood will destroy it. Groups A and B must not be mixed; nor must A, B, or AB be given to people with group O. Group O blood can be given to any other group because it contains no agglutinogens and is therefore known as 'universal donor'. Group AB can receive blood from any other group, because neither A nor B are foreign to people with this group; AB is therefore known as 'universal recipient' (see Table 6·2).

You will notice that group O blood can be given to all groups although it may contain anti-A and anti-B agglutinins. However, the amount of agglutinin donated is too small to have any real effect on the recipient.

Dangers of incompatible blood transfusions

If blood from an incompatible donor is given to an individual, the incoming blood cells are agglutinated by the recipient's agglutinins. This leads to their destruction by white cells. The agglutinated blood may also block the recipient's capillaries, leading to local failure of the circulation. Blockage of capillaries in the heart or brain may even lead to death.

Determination of blood groups is carried out by mixing a few drops of blood separately with group A serum and group B serum. The results are interpreted as in Table 6·3.

Table 6.3 Determination of blood group

If agglutination occurs with	The donor's blood group is
Group A serum only	Group B
Group B serum only	Group A
Group A and Group B serums	Group AB
Neither group A serum nor group B serum	Group O

Rhesus factor is a different blood group system of great medical importance. Up to 85% of white people carry an agglutinogen called the Rhesus factor on their red cells and are therefore known as Rhesus positive (Rh+). The remaining 15% lack this factor and are called Rhesus negative (Rh−). There is no naturally occurring agglutinin against Rhesus factor, but injection of Rh+ blood into an Rh− individual will stimulate production of the anti-Rh agglutinin.

Rhesus factor chiefly affects marriages between Rh− mothers and Rh+ fathers, when it can lead to the illness or death of their second and subsequent children at birth. This occurs because the children are usually Rh$^+$(p154). If some of the Rh$^+$ blood from the foetus leaks across the placenta into the mother's circulation, the agglutinin anti-Rh is manufactured in her blood. If this later leaks back into the foetal circulation, it may agglutinate the embryo's Rh$^+$ blood.

Usually insufficient agglutinins are manufactured to affect the first child, but later children will be affected by the progressive build up of anti-Rh agglutinin. When expectant mothers are known to be Rh$^-$, the husband's blood is also tested. If it is found to be Rh$^+$, the baby may be given a complete transfusion with Rh$^-$ blood as soon as, or even before, it is born.

Transfusions of Rh$^-$ to Rh$^+$ blood are always safe and Rh$^+$ blood can initially be given safely to Rh$^-$ men. However, a second transfusion of Rh$^+$ blood to Rh$^-$ men may result in agglutination of the donated Rh$^+$ cells owing to build up of anti-Rh agglutinin from the previous transfusion.

Rh$^-$ women should not be given Rh$^+$ blood since it will sensitize them to Rh$^+$. This may cause a build up of agglutinin leading to difficulty even with the first-born child.

The circulatory system

The heart

The heart is essentially a large muscular pump, situated in the thorax, under the sternum. It is enclosed in a tough protective membrane called the **pericardium**: the presence of fluid between pericardium and heart reduces friction between the heart and surrounding organs.

The heart is basically two separate pumps joined together and working in unison. Blood cannot circulate between the two sides owing to the presence of a tough muscular **septum** (Fig. 6.2).

Heart beat is caused by the alternate contraction and relaxation of the four muscular chambers of the heart, as follows:

1 Diastole (relaxation phase) The atria (auricles) and ventricles relax (Fig. 6.3a). Deoxygenated blood from the body enters the right atrium through the venae cavae. Oxygenated blood from the lungs enters the left atrium through the pulmonary veins.

2 Systole (contraction phase) The atria contract (Fig. 6.3b), forcing blood into the ventricles, and the ventricles then contract to force blood out of the heart. Blood from the right ventricle is pumped

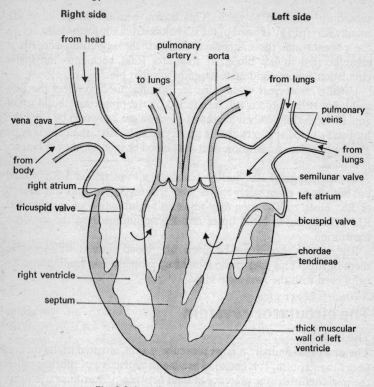

Fig. 6.2 Longitudinal section of the heart

(Note: The heart is always drawn as if facing you. This explains why its left is on your right.)

along the pulmonary artery to the lungs, while blood from the left ventricle is pumped along the aorta to the organs of the body generally.

Action of the valves

Valves prevent back flow at each stage:

1 **Valves in the pulmonary veins and venae cavae** prevent backflow when the atria contract.
2 **The tricuspid and bicuspid (mitral) valves** prevent backflow into the atria when the ventricles contract. Non-elastic tendons, called **chordae tendineae** prevent the valves from being forced back into the atria.

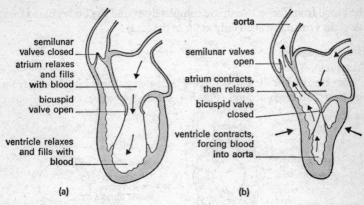

Fig. 6.3 (a) Diastole, (b) systole

3 **Semi-lunar valves** in the aorta and pulmonary artery prevent blood from flowing back into the ventricles when these expand during diastole.

Coronary heart disease

Heart muscle receives its own blood supply through the coronary artery. This artery may become narrowed owing to the deposition of fats on the internal lining, leading to a partial blockage. A blood clot may then become lodged in a branch of the coronary artery, blocking the flow of blood and causing a heart attack or **coronary thrombosis**. As a result, part of the ventricular muscle is unable to function owing to lack of oxygen. In serious cases, this may lead to death. Blockage of the coronary artery is associated with genetic factors, lack of exercise, obesity, an excessively fatty diet, smoking, nervous tension, and high blood pressure.

Blood pressure

The pumping action of the heart maintains a high pressure in the circulatory system in order to force the blood around the body at speed. Pressure is also needed to overcome the resistance to the flow of blood as it passes through the narrow capillaries.

The ventricles are the major source of pressure, especially the left ventricle. The walls of the left ventricle are much thicker than those of the right. The left ventricle is therefore able to contract with greater force, generating a pressure equal to 120 mm of mercury, compared to only 15 mm from the right ventricle. This is necessary in order to force

the blood from the left ventricle completely round the body; blood from
the right ventricle travels only as far as the lungs.

Blood vessels
Arteries

These are thick-walled structures which always carry blood away from
the heart. In cross-section (Fig. 6.4) the artery wall contains an outer

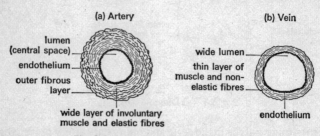

Fig. 6.4 Cross-sections of blood vessels

fibrous protective layer, a middle layer composed of involuntary circular
muscle and elastic fibres, and a thin inner lining, the endothelium.
The lumen, or space in the centre of the tube, is narrower in arteries
than in veins.

The larger arteries, which are nearer the heart, contain extra elastic
fibres to absorb the pressure surge caused by the contractions of the
ventricles. The stretched elastic tissue then contracts between heart
beats so helping to push the blood along the arteries.

There are no valves in the arteries (except near the heart) because the
high blood pressure and squeezing action of the elastic tissue maintains
a steady one-way flow. The cutting of an artery leads to massive loss of
blood as the blood spurts out under pressure.

The arteries branch and become smaller as they leave the heart,
finally dividing into **arterioles**. These contain a relatively thick layer
of circular muscle which contracts or relaxes to control the volume of
blood entering a particular organ.

Capillaries

These are tiny blood vessels, with walls only one cell thick. These thin
walls are permeable, allowing water and dissolved substances such as
oxygen, carbon dioxide, hormones, urea, dissolved food, etc., to pass in
and out of the capillaries. White cells can also squeeze out between
the cell walls.

Veins

Blood from capillaries runs into minute **venules** and then into **veins**. Veins always carry blood to the heart. Compared to arteries, they have a wider lumen and thinner walls, containing non-elastic fibrous tissue instead of the elastic tissue found in arteries.

The blood pressure in veins is steadier and lower than in the arteries. Semi-lunar valves (Fig. 6·5) are present in a number of veins to prevent backflow.

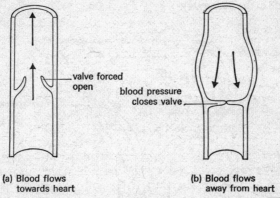

(a) Blood flows
towards heart

(b) Blood flows
away from heart

Fig. 6.5 Longitudinal sections of the veins to show action of the valves

Blood pressure in the veins is maintained by:

1 the remaining effects of arterial pressure;
2 the action of contracting muscles pressing on the veins during exercise;
3 the action of the valves in preventing backflow.

The blood in the veins leaving organs such as the liver, muscles, and brain, generally contains relatively little oxygen and food. It carries more carbon dioxide and nitrogenous waste than the arterial blood entering the same organs. This is because active organs (e.g. the liver) remove oxygen and food substances for use in cell respiration and protein synthesis, and produce carbon dioxide and nitrogenous compounds as waste products.

Special Cases

Those often asked about in exam questions include:

1 The pulmonary artery carries deoxygenated blood from the heart to the lungs.

2 The pulmonary vein carries oxygenated blood from the lungs to the heart.

3 The hepatic portal vein carries a high concentration of glucose and amino acids from the intestines to the liver after a meal.

4 The hepatic vein also carries a high, but more constant, concentration of glucose and amino acids from the liver into the general circulation (p94).

5 The hepatic vein also carries the highest concentration of urea (urea is made in the liver, p94).

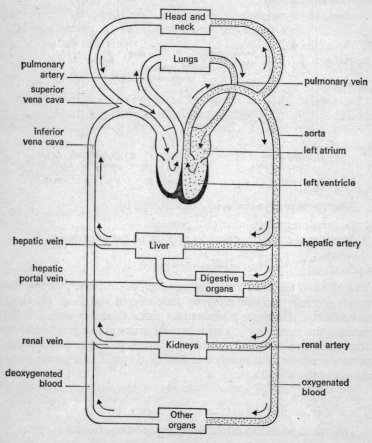

Fig. 6.6 The circulatory system

6 The renal veins carry blood from the kidneys with reduced
   concentrations of urea, salts, and water (p97).
7 The warmest blood is found in the hepatic vein when the body is at
   rest – the liver is a major source of heat.
8 The blood in arteries (except the pulmonary artery) is brighter red,
   owing to the presence of oxyhaemoglobin. Venous blood (except for
   the pulmonary veins) is darker red, owing to the presence of
   haemoglobin without oxygen.

Foetal circulation

The **foetal heart and circulation** differs from the adult arrangement
because the foetus receives its oxygen supply from the placenta (p134).
The lungs are not used until birth, and are by-passed in the following
ways (Fig. 6.7).

1 A gap called the **foramen ovale** ('oval hole') exists in the septum
   between the atria. This allows most of the blood to flow directly from
   the body, through the right atrium, into the left atrium, and out via
   the left ventricle to the body.
2 A small amount of blood passes into the right ventricle, but a
   connection between the pulmonary artery and aorta (the **ductus
   arteriosus**) allows much of this blood to pass into the aorta instead
   of the lungs.

At birth the foramen ovale between the atria usually closes within a
week and the ductus arteriosus closes in a few hours, so directing the
blood through the lungs. The condition known as a 'hole in the heart'
occurs when the foramen ovale fails to close.

Exchange of substances between capillaries and tissues

As blood passes through the capillaries, oxygen and dissolved food
substances are lost from the blood to the surrounding tissue fluid while
carbon dioxide and nitrogenous wastes are taken in. **Tissue fluid** is a
clear liquid similar to plasma in composition, but lacking red cells and
containing a lower concentration of proteins. It surrounds the cells and
acts as a medium for the exchange of substances with the blood.

The exchange of substances between the capillaries and the tissue
fluid begins when blood enters the capillaries from the arterioles, under
pressure. The high blood pressure at the arterial end forces fluid
through the capillary walls (Fig. 6.8). Towards the venous end, blood
pressure falls. Fluid is absorbed into the capillaries because the osmotic

pressure of the plasma is greater than both the reduced blood pressure
at the venous end and the osmotic pressure of tissue fluid.

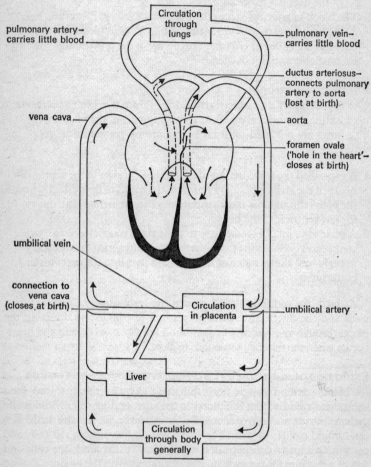

Fig. 6.7 Circulation in the foetus

Oxygen, glucose, and amino acids diffuse from the tissue fluid to the
cells, while carbon dioxide and nitrogenous wastes diffuse from the cells
into the tissue fluid.

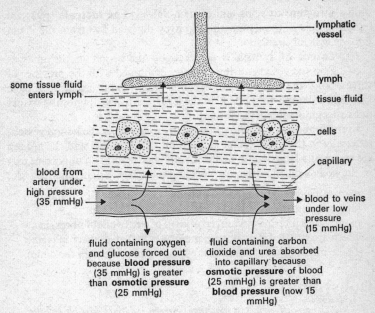

Fig. 6.8 Exchange of substances between capillaries and tissues

(**Note:** Pressure is also measured in pascals, abbreviated to Pa. 35 mmHg ≃ 4600 Pa; 25 mmHg ≃ 3300 Pa. You do not need to know these units, but you may need to recognize them in examination questions.)

Lymphatic system

A second system, the lymphatic system, is available to return excess tissue fluid from the tissues to the main circulatory system. It also plays a major role in the body's defence system.

Minute lymph vessels permeate the tissues in a similar manner to capillaries, except that lymphatic capillaries end blindly. Surplus fluid drains into these vessels and passes back to the main veins in the form of lymph.

Lymph is a fluid similar in composition to tissue fluid, but containing extra lymphocytes since these are manufactured within the lymphatic system. Lymph is also similar in composition to plasma but it contains a lower concentration of protein as well as extra lymphocytes. Red cells are never present in lymph. The lymph vessels join together to form two main **lymphatic ducts**. These return fluid to the main veins through openings in the subclavian veins below the neck. Lymph vessels have a similar structure to veins and contain valves. **The lacteals** (p93) are lymph vessels in the ileum which convey digested fats from the ileum into the general circulation.

Movement of lymph is achieved by the action of the muscles pressing on the vessels during exercise, combined with the presence of valves, and the pressure of accumulating lymph itself. Complete blockage of lymph vessels, e.g. in the leg, leads to swelling of the tissues due to accumulation of tissue fluid.

Lymph nodes are swellings found in groups along the lymph vessels, especially in the neck, groin, and armpit. The **tonsils** and the **spleen** are essentially large lymph nodes. Lymph nodes form an important part of the body's defences against invading microbes. Their functions are:

1 to manufacture lymphocytes, which produce antibodies against incoming antigens.
2 to provide a filtering system, composed of a network of fibres, in which are situated stationary granulocytes known as macrophages. These ingest foreign bodies as the lymph passes through the nodes.

Questions

1 *Give* **three** *differences in structure between red and white blood cells.* [p58; 6]

Answer guide
When answering questions involving comparisons, it is essential to mention both halves of the comparison in the same sentence. For example: 'White cells have a nucleus but red cells lack nuclei'. Do not write down a list of facts about red cells, followed by a separate list of facts about white cells.

2 *State* **three** *causes of anaemia.* [p58; 3]
3 *What are the differences between each of these pairs?*
   (a) *antigens and antibodies*
   (b) *natural and artificial immunity*
   (c) *active and passive immunity.* [p61; 2 each]
4 (a) *Explain how agglutinogens (antigens) and agglutinins (antibodies) react together to produce the blood groups A, B, AB, and O.* [p63; 7] (AEB)
   (b) *Describe the procedures by which the A, B, AB, and O blood groups can be determined.* [7]
5 *Write an illustrated account of the pumping action of the heart.* [p66; 10]

6 *How does foetal circulation differ from that of the adult? Describe and explain the changes in the foetal circulation at birth.* [p71; 6]

7 *By means of fully labelled transverse sectional diagrams, show the differences in structure between arteries, veins, and capillaries.* [p68; 7] (WJ)

   Draw and label fully a longitudinal section of a vein to show clearly any other structures present. [2]

   Explain fully how the structure of each vessel is adapted for the function it performs. [7]

8 (*i*) *Make a clear labelled diagram to show the possible route taken by a red blood corpuscle as it leaves the right ventricle until it reaches the liver.* (A detailed drawing of the heart is *not* required.) [p70; 10] (OX)

   (*ii*) *Describe the changes which take place in the red blood corpuscle and in the surrounding plasma during the journey.* [p58; 15] (OX)

Answer guide

(i) The diagram will contain the relevant parts of Fig. 6.6 on p70 [10].

(ii) On entry into the lungs, the corpuscle takes up oxygen [1] to form oxyhaemoglobin [1].

   Carbon dioxide diffuses out of the plasma [1]. After passing through the heart, some of the blood enters the liver directly, but some passes into the capillaries surrounding the stomach and intestines. Here the oxyhaemoglobin in the red corpuscles breaks down to form oxygen [1] and haemoglobin [1] and the oxygen diffuses out of the blood into the tissues [1]. Carbon dioxide diffuses out of the tissues into the plasma [1].

   Dissolved food substances, e.g. glucose [1], amino acids [1], and some fatty acids [1] and glycerol [1] enter the plasma from the tissues. On entry into the liver, the blood in the hepatic portal vein [1] therefore contains a high concentration of dissolved food substances [1] and carbon dioxide, [1] and a low concentration of oxygen [1].

9 *The blood passing through the aorta as it leaves the heart has a high pressure and flows unevenly. In the small arterioles the pressure is lower and the flow smoother.*

   Explain the part played by elastic and muscular tissue in relation to these changes. [p68; 5] (AEB)

10 (*a*) *Distinguish between plasma and lymph.* [p73; 4]

   (*b*) *Give a concise account of the lymphatic system and describe its roles as*
   (*i*) *a transport medium;*
   (*ii*) *a defence system of the body.* [p73; 16], (CAM)

7 Nutrition

Food is needed to provide raw materials for growth and replacement of cells, and to provide energy. The four major components of a healthy diet are proteins, fats, carbohydrates, and water.

Food components

Proteins are organic chemicals which always contain carbon, hydrogen, oxygen, and nitrogen; they may also contain sulphur and phosphorus. Proteins are long molecules, consisting of various combinations of amino acids.

Amino acids are joined together by means of a chemical link called a

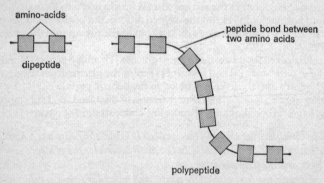

Fig. 7.1 Structure of protein

peptide bond. Two amino acids joined together by a peptide bond are known as a **dipeptide**; a molecule composed of many amino acids is called a **polypeptide** (*see* Fig. 7.1). Amino acids are built up into proteins in the cytoplasm according to coded instructions in the nucleus (p19). Proteins cannot be stored in the body and so a steady supply of amino acids is needed from the diet.

Proteins are the body's main building blocks. They are the main component of cell membranes, and of cytoplasmic structures such as mitochondria, endoplasmic reticulum, and ribosomes (p20). Muscles contain a high concentration of protein, and proteins are also found in bone.

Carbohydrates (e.g. sugar and starch) are organic chemicals containing carbon, hydrogen, and oxygen only. There is always twice as much hydrogen as oxygen; for example glucose, $C_6H_{12}O_6$; and sucrose (cane sugar), $C_{12}H_{22}O_{11}$. Carbohydrates are composed of a series of rings

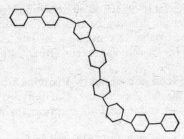

(a) Monosaccharide, e.g. glucose, $C_6H_{12}O_6$

one ring

(b) Disaccharide, e.g. maltose, $C_{12}H_{22}O_{11}$

two rings joined together

(c) Polysaccharide, e.g. starch

several hundred rings joined together

Fig. 7.2 Structure of carbohydrates

(Fig. 7.2), each containing six carbon atoms with attached hydrogen and oxygen atoms.

Simple sugars, such as glucose, contain one ring only and are called **monosaccharides**. More complex sugars, such as maltose, $C_{12}H_{22}O_{11}$, are made up from two glucose molecules joined together. Maltose, sucrose, and other C_{12} sugars are called **disaccharides**; larger carbohydrates, known as **polysaccharides,** are made from several hundred glucose molecules.

Polysaccharides are insoluble and not sweet, unlike the mono- and disaccharide sugars. Examples include starch, used as a common storage material in plant cells, and cellulose, found in plant cell walls. Animals cannot make starch, but use a polysaccharide called **glycogen** as a storage carbohydrate.

Carbohydrates are needed mainly as a source of energy to drive all of the chemical processes occurring in living cells. Energy is obtained by breaking down glucose during tissue respiration (p21). Surplus glucose molecules are combined to form glycogen, which is built up from thousands of glucose molecules. Glycogen is mainly stored in the liver and the muscles.

Fats and oils are also organic chemicals containing carbon, hydrogen, and oxygen. They are composed of one molecule of **glycerol** and three molecules of **fatty acid** joined together. Oils are essentially fats which happen to be liquid at room temperature. Fats and oils are together called **lipids**. Lipids are found in cell membranes and the many cytoplasmic structures composed of folded membranes (e.g. mitochondria and endoplasmic reticulum).

Fats are also used in the body as a store of energy. They are suited to this purpose because they are insoluble and yield twice as much energy per gram as carbohydrates. Fats are also excellent insulators; the layer of subcutaneous fat (p29) acts as an insulating layer to retain body heat. Many, but not all, fats can be manufactured in the body from carbohydrates. It is essential to consume some fats, especially as these are a good source of the fat-soluble vitamins (A, D, and K).

Sources of major food components **Proteins** are obtained chiefly from meat, eggs, fish, nuts, and cheese. **Carbohydrates** are found in potatoes, bread, cereals, and sugar. **Fats** are present in butter, margarine, cheese, and fatty meat.

Vitamins are organic chemicals required in minute quantities for certain essential chemical activities in the body. None can be manufactured by the body (except vitamin D); all must be present in sufficient quantities in the diet. If the diet is lacking in a particular vitamin, characteristic symptoms of disease appear. These are known as **deficiency diseases** (see Table 7.1). If the missing vitamin is later supplied, the symptoms rapidly clear up.

The importance of vitamins was first demonstrated by **Gowland Hopkins** in 1912. He fed a group of rats (group A) on a diet composed of pure carbohydrates, fats, proteins, mineral salts, and water. They hardly grew at all for 18 days, while a second group (B) on the same

Table 7.1 Importance of vitamins in the diet

Name	Main sources in the diet	Deficiency diseases (names in italics)
A	Milk and butter, cod liver oil, liver, and fresh green vegetables	Poor resistance to infections; poor night vision; dry scaly skin (*Night blindness*)
B$_1$ (thiamine)	Yeast and wheat germ. White flour (wheat germ removed) and polished rice both lack vitamin B$_1$; brown flour and rice husks are good sources.	Loss of appetite, leading to loss of weight; feet and legs may swell (*beri-beri*)
Other B vitamins	Milk, meat, and green vegetables	Various signs of ill-health. Shortage of B$_6$ (nicotinic acid) leads to diarrhoea, loss of weight, and mental disorders (*pellagra*). Lack of B$_{12}$ is a cause of *anaemia*.
C (ascorbic acid)	Fresh fruit and vegetables especially oranges, lemons, tomatoes, and blackcurrants; potatoes and liver also contain moderate amounts	Capillary bleeding, skin disease, poor resistance to infections, slow healing of wounds (*scurvy*)
D	Cod liver oil, cream, and egg yolk; can be manufactured in the skin during sunlight	Fragile, soft bones with swollen ends; leg bones often bend out of shape (*rickets*). Essential for the absorption of calcium and phosphate from digested foods.
K	Vegetables, e.g. spinach; also synthesized in the gut by bacteria	Clotting time of blood is increased, leading to slow healing. Other symptoms similar to scurvy.

diet, but with 3 cm^3 of milk added daily, grew rapidly. After 18 days he transferred the milk from group B to group A. Two days later, group A began to grow rapidly (Fig. 7.3) while group B's growth slowed down.

Evidently the milk contained small quantities of chemicals essential to growth. (Questions on this experiment are common in exams.)

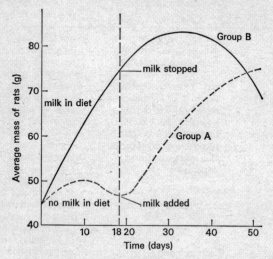

Fig. 7.3 Results of Gowland Hopkins' experiment

Mineral salts Various elements are essential in small quantities for health (*see* Table 7.2).

Water is essential for all chemical reactions in the body. Lack of water leads to death within three days, since the cells die from dehydration. Water is especially required to replace perspiration, and for digestion and transport of dissolved food, etc., in the blood. Two to three litres are lost each day in urine, faeces, and evaporation from the lungs and skin. One and a half litres daily are obtained by drinking liquids, and the remainder from the respiration of food (p21).

Dietary fibre (roughage) is the name given to those constituents of the diet which pass through the body undigested. These consist mainly of cellulose fibres from plants found in bran, unrefined (wholemeal) bread, and fruit and vegetables. Fibre adds bulk to the faeces and so stimulates peristalsis (p90) in the colon. Lack of fibre may cause constipation.

Balanced diet
A balanced diet contains sufficient food in total to provide an adequate supply of energy for daily activity; proteins, fats, and carbohydrates in

Table 7.2 Mineral salts needed for health

Name	Sources	Effects of deficiency
Iron	Liver, eggs, spinach, and other vegetables	Needed to make haemoglobin, so shortage may cause anaemia (lack of red cells)
Sodium	Table or cooking salt (sodium chloride)	Essential component of plasma; without it the salt balance of the body is upset, leading to muscular cramp
Chlorine	Table or cooking salt (sodium chloride)	As for sodium; also used to make hydrochloric acid in the stomach
Potassium	Cereals	Essential for growth, especially of muscles and red cells
Iodine	Occurs as iodide in tap water and sea foods; may need to be added to water as iodized salt	Required for the manufacture of the hormone thyroxine (p111); absence leads to goitre or cretinism
Calcium	Cheese, milk, beans, and tap water in hard water areas.	Required for clotting of blood and conduction of nerve impulses; combined with phosphate in bones and teeth; shortage may lead to rickets
Phosphorus	Occurs as phosphate in meat, eggs, fish	As for calcium in bones and teeth; also needed to make ATP (p21) and nucleic acids (p23)
Fluorine	Occurs as fluoride in some natural water supplies	Essential for the growth of healthy teeth and prevention of tooth decay (p86)

the correct proportions; vitamins, minerals, salts, water, and fibre all in sufficient quantities.

Milk is almost a complete food in itself since it contains protein, fat, carbohydrate, minerals, especially calcium and magnesium, and vitamins A, B, C, and D. The only important substance missing is iron but the baby is usually born with sufficient iron in its body for the first weeks of life.

Energy value of food

The total amount of food consumed each day must provide sufficient energy to keep the body alive and active (*see* Table 7.3). Energy is chiefly required to replace heat lost to the outside.

Metabolic rate This is the rate at which the body uses up energy. The rate varies partly according to age and sex and partly according to activity.

The **basal metabolic rate** is the rate at which energy is used up simply to maintain life, i.e. for breathing, heartbeat, growth and repair, and to replace lost heat. The basal rate is high at birth and falls slowly with increasing age; men have a higher basal rate than women.

A typical European male requires about 7500 kilojoules (kJ) daily merely to maintain basal metabolism. The overall metabolic rate equals the basal rate plus the extra kilojoules required for daily activities. Even the act of sitting up increases the demand for energy, compared to lying in bed.

Table 7.3 Typical daily energy requirements

By age	Requirement (kJ)
6-month-old-baby	5 000
5-year-old-boy	7 250
16-year-old-boy	12 000
Inactive 25-year-old-man	10 500
75-year-old man	9 000
By sex and activity	
16-year-old girl	9 000
16-year-old boy	12 000
Moderately active man aged 25	12 000
Manual labourer aged 25	14 000

Young children require fewer kilojoules than adults simply because they have smaller bodies. They consume more food than adults in proportion to their size, owing to their higher basal metabolic rate, and their steady growth. They also need more protein in proportion to the rest of their diet in order to sustain growth.

Teenagers require more food than adults following similar activities. Extra kilojoules and more protein are required to sustain the rapid growth which follows adolescence.

Old people need fewer kilojoules overall, as their metabolic rate is lower than average; and they are not growing.

Inhabitants of cold countries obviously require more kilojoules than tropical dwellers. Fat is especially valuable for Arctic inhabitants because of its high energy content.

Questions

1 (a) List four elements always found in proteins. [p76; 4]
  (b) Name one other element present in some proteins. [1]
  (c) Name the basic units from which proteins are made. [1]
2 Why is protein necessary in the diet? [p77; 3]
3 Give three main sources of (a) protein, (b) carbohydrates, in the diet.
  [p78; 6]
4 What are vitamins? [2]
5 Give three sources each for vitamins B and C in the diet. [p79; 6]
6 (a) Why is milk regarded as almost a complete food? [p81; 1]
  (b) Which component of a balanced diet is lacking from milk? [p81; 1]
7 What is meant by a 'balanced diet'? [p80; 4]
8 At what stages in the human life cycle is an adequate supply of protein
  particularly needed? [p82; 2]
9 Describe, with reasons, the differing nutritional needs of [p82; 3 each]:
  (a) an Arctic inhabitant; (b) a five-year-old boy; (c) a farm worker;
  (d) a pregnant woman; (e) a teenage girl.
10 A shift worker in a slate quarry works for 8 hours a day on heavy
   manual work. He sleeps for another 8 hours and is active for the other
   8 hours. His basic rate of energy use is 300 kJ per hour; for light
   activity 500 kJ per hour; for heavy work 1500 kJ per hour.
   What are his total energy needs for the day? Show your working.
   [p82; 5]   (MREB)

Answer guide
   While asleep, he uses 8 × 300   = 2400   kJ
   For light activity, he uses 8 × 500 = 4000   kJ
   For heavy work, he uses 8 × 1500 = 12 000 kJ
∴ His total energy need   = 18 400 kJ

8　Digestion

Digestion is the breakdown of large insoluble molecules to small soluble molecules capable of absorption into the bloodstream. Most of the molecules in food are too large to pass through the walls of the gut without digestion. Digestion involves both the **physical** breakdown of food (e.g. by the teeth) and its **chemical** breakdown (e.g. by the action of water).

Digestion in the mouth

Digestion begins as soon as food enters the mouth, by mechanical chewing and by the secretion of saliva. The function of the teeth is to break food into small pieces, and the function of saliva is to moisten the food (p87).

Teeth

Teeth are outgrowths from the skin covering the upper and lower jaws. All teeth have the same basic structure.

Structure of a generalized tooth

Teeth consist of a crown and root. The outer layer of the crown is composed of hard, non-living **enamel**, consisting mainly of mineral salts, chiefly calcium phosphate. Enamel forms a hard biting surface, and also protects the layers beneath it.

Dentine The layer of living material found below the enamel is called dentine. It is similar to bone in composition and therefore softer than enamel. Dentine makes up the bulk of each tooth.

Pulp cavity The space inside the tooth is filled by a soft connective tissue called **pulp**. This contains sensory nerve fibres, and capillaries, both of which penetrate the dentine to some extent. The capillaries supply food and oxygen to cells in the dentine.

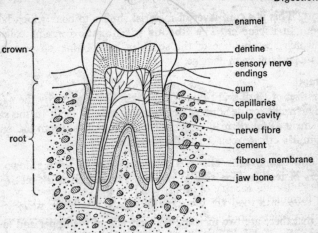

Fig. 8.1 Vertical section through a molar tooth

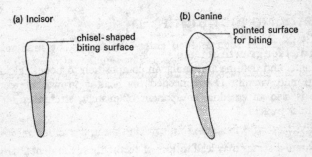

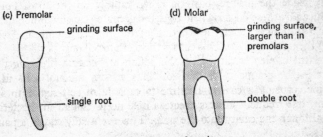

Fig. 8.2 Types of teeth

Cement This is a thin layer of material similar to bone which holds the tooth in its socket. The **fibrous membrane** contains collagen fibres which absorb some of the shock of hard biting, so protecting tooth and jawbone from damage.

Milk teeth

The first set of teeth erupt through the skin of the gums by the age of two. This milk, or deciduous, set contains twenty teeth in the same arrangement as the permanent set but without molars. From the age of six the milk teeth are replaced by the permanent set.

The dental formula This is a shorthand way of writing down the numbers of teeth in one side of the jaw. For human **milk teeth**, the dental formula is incisors $(\mathbf{I})\frac{2}{2}$, canines $(\mathbf{C})\frac{1}{1}$, Premolars $(\mathbf{P})\frac{2}{2}$. This means that there are two incisors, on each side of the upper and lower jaws; together with one canine and two premolars. The dental formula for the permanent dentition is written as:

$$I\frac{2}{2} \quad C\frac{1}{1} \quad P\frac{2}{2} \quad M\frac{3}{3}$$

Correct diet for growth of teeth

Since enamel and dentine are similar in composition to bone, calcium phosphate and vitamin D are needed for healthy growth of teeth. Fluoride is also an essential component of enamel, strengthening it against attack by acid.

Tooth decay

Two separate diseases may lead to loss of teeth: these are dental caries (tooth decay) and periodontal (gum) disease.

Periodontal disease is caused by an accumulation of a soft, sticky material known as **plaque** on the teeth. Plaque is a mixture of saliva and food; inside the plaque, bacteria breed rapidly on the food. Poisons produced by these bacteria penetrate the gums, so causing inflammation of the gums and, later, destruction of the fibrous membrane which holds the teeth in place. This causes the teeth to fall out.

Dental caries is caused by the action of bacteria breaking down sugar in the plaque to form acid. This attacks mineral salts in the dentine, causing a cavity. Dentine is capable of self-repair to some extent, but frequent attacks create a hole deep enough for bacteria to pass through the cavity into the pulp. This eventually causes a painful abscess.

Prevention of periodontal disease and dental caries is best achieved by regularly removing plaque and preventing its accumulation. The major points to remember are:

1 **Sugar,** and foods containing sugar (e.g. cake, biscuits, soft drinks), should only be eaten at mealtimes. Frequent consumption of sugar prevents self-repair by the dentine.
2 Plaque should be removed regularly by **brushing** hard across the teeth. Thorough, daily brushing is very important; but quick, frequent brushing is ineffective.
3 **Fluoride** is essential for healthy teeth. If it is not available in the water supply, fluoride tablets can be taken. Brushing with a fluoride toothpaste is helpful.
4 Regular visits to the **dentist** are necessary to help improve badly shaped teeth. These may trap plaque in places which are difficult to brush. The dentist will also fill cavities and remove deposits of hardened plaque known as calculus (tartar) from the teeth.

Note: Much of the information given in older textbooks on prevention of tooth decay is incorrect.

Mastication (chewing)
After food has been bitten off by the incisors, it is chewed by the premolars and molars. Chewing breaks food into smaller pieces; this allows saliva to become mixed with it, and increases the surface area available for digestion by enzymes. It also makes the food easier to swallow.

Saliva is mainly composed of watery mucus, salts and the enzyme **salivary amylase** (ptyalin). Salivary amylase breaks down starch into maltose sugar (p77). Saliva also contains antibodies and an enzyme called **lysozyme** which breaks down bacterial cell walls. This helps prevent infection by microbes in food.

The mucus in saliva lubricates the food, which makes swallowing easier, and also dissolves it slightly. Dissolved food can then be tasted by the tongue. Saliva is secreted from three pairs of **salivary glands** situated under the tongue and at the back of the mouth. The flow of saliva is stimulated by seeing, smelling, tasting, or even thinking about food.

Swallowing
Chewed food particles are collected into a ball, called a **bolus,** ready for swallowing as follows:

1 **The tongue** pushes backwards and upwards against the **hard** palate (*see* Fig. 5.1), so forcing the bolus to the back of the mouth.

2 The opening between the nasal cavity and the pharynx is closed by the **soft palate**.

3 The **laryngeal cartilages** move upwards by muscle action. As a result, the larynx constricts and becomes sealed off under the **epiglottis**.

4 The bolus is pushed into the oesophagus and passed down to the stomach by **peristalsis** (p90).

Digestion in the stomach

Functions of the stomach

Storage

As the stomach is a muscular sac with flexible walls, large quantities of food can be stored for several hours. Food is released from the stomach at regular intervals by the opening of the **pyloric sphincter**. This is a tight ring of muscle used to retain food in the stomach.

Mechanical digestion

While in the stomach, the food is churned by muscular contractions of the stomach wall, and converted into a semi-liquid state by being mixed with gastric juice.

Chemical digestion

At the sight, smell, or expectation of food, the gastric glands in the stomach wall secrete **gastric juice**. This contains water and mucus to dissolve and lubricate food; the mucus also protects the wall of the stomach from the effects of enzymes.

Gastric juice also contains dilute hydrochloric acid and the enzymes pepsin and rennin. **Pepsin** is secreted in an inactive form called **pepsinogen,** so preventing digestion of the gastric gland cells. In the presence of hydrochloric acid, pepsinogen is converted to pepsin. Pepsin acts on proteins, breaking them down into peptides.

Rennin acts on milk, converting soluble milk proteins into an insoluble form. This causes the milk to be retained in the stomach for enough time for the pepsin to act on the proteins. Rennin is found chiefly in babies.

The **hydrochloric acid** in the gastric juice provides the optimum pH for the action of pepsin and rennin; kills various bacteria present in food; and converts pepsinogen to pepsin.

Structure of the stomach

The stomach consists of a large, muscular, elastic bag. In cross section (Fig. 8.3), layers of muscle fibres can be seen. These carry out the regular churning motions.

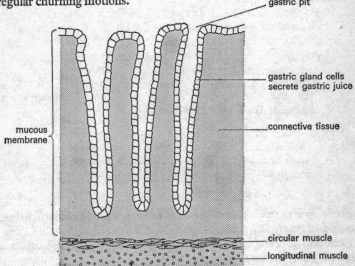

Fig. 8.3 Cross-section of stomach wall

The tissue between the muscles and the interior of the stomach is called the mucous membrane. It contains mucous gland cells, connective tissue, and a thin, moist epithelium. In the stomach, the mucous membrane contains numerous glands inside deep pits. Cells in these glands secrete pepsinogen, acid, rennin, and mucus.

Digestion in the intestines

Digestion in the duodenum

The duodenum is a short curved tube leading from the stomach to the ileum. Its mucous membrane cells produce both mucus and an enzyme called **enterokinase**. As food passes through the duodenum, secretions from the gall bladder and pancreas are also poured onto it.

The gall bladder acts as a storage organ for bile, which is produced in the liver. **Bile** is a green, alkaline liquid which is released through the bile duct whenever food enters the duodenum. The salts contained in bile act on fat droplets, lowering their surface tension and causing them

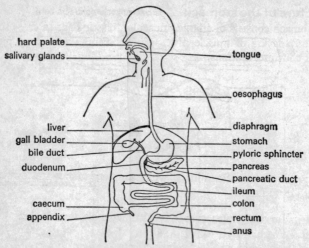

Fig. 8.4 The alimentary canal

to break up into smaller droplets. This process, called **emulsification**, creates a larger surface area for the action of the enzyme lipase.

Pancreatic juice contains water, sodium hydrogen carbonate, and three enzymes, one for each class of foods:

1 **pancreatic amylase** breaks down starch;
2 **pancreatic lipase** splits fats into fatty acids and glycerol;
3 **trypsin** converts proteins and peptides to amino acids.

Trypsin is secreted as the inactive trypsinogen, and is activated by the enterokinase secreted by the duodenal mucosa. The sodium hydrogen carbonate in the juice neutralizes the stomach acid, so creating an optimum pH for the enzymes acting in the ileum.

Peristalsis
Food is moved through the digestive system by means of muscular contractions of the ring of circular muscles which lines the gut. Continuous waves of contraction cause each bolus of food to move at a rate of about 2 centimetres per second.

Digestion in the ileum
Most of the breakdown and absorption of food occurs within this long tube, which stretches for up to 3 metres. The mucosal lining of the

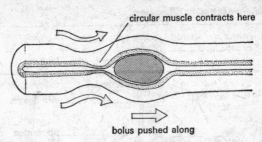

circular muscle contracts here

bolus pushed along

Fig. 8.5 Peristalsis

ileum produces **intestinal juice,** which contains mucus and a variety of enzymes including:

1 **maltase** and other enzymes to break down maltose and other sugars into glucose;
2 **peptidases** (known collectively as erepsin) to break down intermediate protein compounds (p96) into amino acids;
3 **lipase** to break down fats.

As a result, all digestible material is now finally reduced to simple soluble compounds which can pass through the lining of the ileum into the blood. The undigested material, chiefly cellulose from plant cell walls, is passed out through a sphincter muscle into the colon.

The large intestine (colon and rectum)
The food which enters the colon consists chiefly of indigestible plant remains (dietary fibre), dead bacteria, and dead cells, all in a semi-liquid state, in water. The cells lining the colon absorb most of the water from this waste material.

Finally the residue is passed into the rectum, and expelled as the semi-solid faeces through the anus. **Defaecation** (passing out of faeces) takes place at intervals which vary widely from person to person.

The **faeces** consist mainly of dead bacteria and dead cells from the lining of the intestine. Extra bulk can be given to the faeces by eating food containing fibre. This helps to prevent constipation, since peristalsis occurs more easily if the faeces contain sufficient hard material.

The caecum and appendix are both very small in man. Their functions are not fully understood but they contain bacteria which assist in the manufacture of vitamin K and some of the B group vitamins.

Table 8.1   Main stages in digestion

Region of alimentary canal	Name and source of digestive juice	Active constituents of the juice	Action
Mouth	Saliva ; made in the salivary glands	Salivary amylase	Converts starch to maltose
Stomach	Gastric juice ; made in the gastric glands in the wall of the stomach	Pepsin	Converts proteins to peptides ; requires acid environment to function.
		Rennin	Clots proteins in milk ; found mainly in babies
		Hydrochloric acid	Provides acid environment for the action of pepsin and rennin
Duodenum	Pancreatic juice ; made in the pancreas	Trypsin	Converts proteins and peptides to amino acids
		Amylase	Converts starch to maltose
		Lipase	Converts fats to fatty acids and glycerol
	Bile ; made in the liver and stored in the gall bladder	No enzyme present ; contains alkaline salts	Emulsifies fats, i.e. breaks down large droplets
Ileum	Intestinal juice, made in glands in the ileum lining between the villi	Erepsin	Converts peptides to amino acids
		Lipase	Converts fats to fatty acids and glycerol
		Maltase and other enzymes acting on sugar	Convert maltose and other sugars to glucose

Absorption of digested food

In the stomach, only water, salts, and alcohol are absorbed from the food. This explains why alcohol acts so rapidly on the body. In the colon, water only is absorbed. All other absorption takes place in the ileum.

The structure of the ileum is adapted for absorption because its great length creates a **large surface area**, and the surface area of the intestinal lining is also increased by the presence of thousands of tiny projections called **villi**. The epithelium of the villi is very thin so allowing rapid passage of dissolved molecules through it, and each

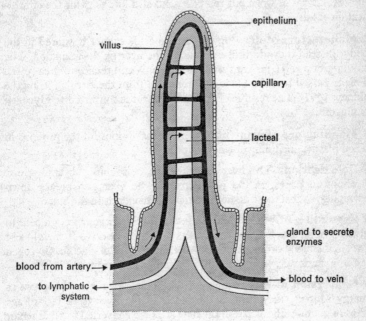

Fig. 8.6 Structure of a villus

villus contains a dense network of **blood capillaries** for rapid transport of amino acids and glucose.

Glucose and **amino acids** are transported in the capillaries of the villi, which unite to form the hepatic portal vein. This carries all of the blood from the ileum to the liver.

Fats are either absorbed as fatty acids and glycerol or as tiny droplets of emulsified fat. The fats enter the lacteals and are carried through the lymph system, finally emptying into the circulatory system near the neck. Fatty acids and glycerol may enter the capillaries directly.

Uses of absorbed food

Most of the food absorbed through the villi passes into the hepatic portal vein and thence to the liver.

Functions of the liver

These include:

1 Manufacture of bile Bile contains waste substances produced by the breakdown of worn out red cells, and also bile salts for the emulsification of fats.

2 Manufacture of urea Excess amino acids cannot be stored in the body; instead they are broken down in the liver by de-amination. The amino group (NH_2) is converted to ammonia and thence to urea (p97). This is passed into the blood and removed from the circulation by the kidneys (p97). The residue of each amino acid is converted to glycogen or glucose.

3 Manufacture of proteins The liver makes most of the proteins in blood plasma, including fibrinogen (p61).

4 Detoxication The liver breaks down poisonous substances, e.g. alcohol and drugs, in the bloodstream. The liver also breaks down hormones, so that these do not circulate permanently in the blood.

5 Regulation of blood glucose level The liver converts glucose to **glycogen,** a carbohydrate which acts as a food reserve. If the glucose level in the blood drops, glycogen is converted back to glucose. Glycogen acts as a quick release store for the production of energy.

6 Conversion of fats Similarly, when fats are required for use as energy sources, they pass in the blood to the liver from fat storage areas (e.g. under the skin). In the liver, fats are converted to glycogen or glucose. Fats therefore act as a slow release store of energy.

7 Storage In addition to glycogen, the liver stores iron (from the breakdown of red cells) and vitamins A, D, and B_{12}.

8 Production of heat The constant activity of the liver cells produces much of the body's heat.

Questions

1 *Make an outline diagram of the overall shape of the four main types of teeth* [p85; 4] *and state the function of each.* [4]

2 *Name one vitamin and three mineral elements necessary for healthy teeth.* [p86; 4]

3 (a) *What are the two main causes of dental decay?* [p86; 2]
  (b) *How can dental decay be prevented?* [4]

4 *Write an account of the stomach, using the following headings as a guide.* (p88)
  (a) *Its shape and position relative to the organs of the abdomen (NB A drawing of the whole alimentary canal is not required).* [5]
  (b) *How it mixes and retains food.* [4]
  (c) *The liquids which it produces and the value of each to the body.* [7] (WJ)

5 *Digestion may be considered as a process of the unlinking of large molecules. What is the importance to the body of this unlinking?* [p85; 2] (AEB)

6 (a) *Describe the stages in the digestion of a piece of cheese.* [p90; 10]
  (b) *What use is made by the body of the products of its digestion?* [4]

Answer plan for part (a):
 (i) Mention action of saliva and teeth.
 (ii) Describe protein digestion.
 (iii) Describe fat digestion.
 (iv) Give a diagram (e.g. Fig. 8.4).
      Note that you will not get any extra marks for mentioning carbohydrates—but you will have wasted valuable time.

7 *How is the structure of (a) the ileum* [p90; 2], (b) *an individual villus* [2], *suited to their functions?*

8 *What is the role of the liver* (p94) *in (a) nitrogen metabolism,* [3] (b) *homeostasis, in relation to the composition of blood?* [12]

Answer guide
(a) This concerns the de-amination of amino acids to ammonia and urea (p94).
(b) The following liver functions should be included: breakdown of worn out red cells, manufacture of blood proteins, detoxication, regulation of blood glucose level. All of these relate to homeostasis (p60).

9 Excretion

Introduction

Excretion is the removal from the body of waste products produced by metabolism. It must not be confused with defaecation (p91), which is simply the passing out of undigested food and other material through the anus. Undigested food never enters the circulation and so does not affect the metabolism. The chief excreted wastes are shown in Table 9.1.

Table 9.1 Excreted waste products

Substances	How produced	How excreted
Carbon dioxide	From cell respiration (p21)	In expired air, from the lungs
Urea	From excess amino acids	In urine, from the kidneys; small amount present in sweat
Water	From cell respiration and from the diet	In urine
Salts	From the diet	In urine and in sweat

Water, salts, and a little urea are also lost in the faeces.

Excretion is chiefly carried out by the kidneys. The kidney also play a major role in maintaining constant concentrations of salts, water, etc., in the body fluids. This is an example of homeostasis (p60).

Osmoregulation

The salt/water balance of the body fluids is also referred to as the osmotic pressure of the fluid. The kidney therefore plays a major role in maintaining constant osmotic pressure. This is known as osmo-regulation.

Origin of urea

Urea is produced from the breakdown of excess amino acids in the liver (p94). These are deaminated in the liver to produce ammonia (NH_3). **Ammonia** is highly poisonous but it is rapidly converted to urea in the liver. Urea is less poisonous than ammonia and is very soluble in water. It passes out of the liver into the bloodstream, through the hepatic vein, and is subsequently extracted from the blood by the kidneys.

Kidney structure

The kidneys are a pair of bean-shaped organs situated in the abdomen (*see* Fig. 9.1) between the muscles of the lower back, and the digestive

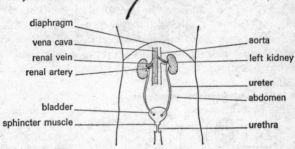

Fig. 9.1 Position of kidneys in the body

organs, for protection. The **renal arteries** supply oxygenated blood to the kidneys and the **renal veins** remove deoxygenated blood.

The kidneys are connected to the **bladder** by means of the **ureters**. The bladder is drained by the **urethra**, which opens directly to the outside in females, and through the penis in males (Fig. 12.1). Seen through a hand lens (Fig. 9.2), the kidneys are composed of a dark outer **cortex**, a lighter inner **medulla**, and the wide mouth of the ureter, called the **renal pelvis**. Extensions of the medulla called **pyramids**, project into the pelvis.

Internal structure

The kidney is largely composed of a mass of tubules called **nephrons** (Fig. 9.3). Each nephron begins with a cup-shaped structure called a **Bowman's capsule** in the cortex. This encloses a tightly coiled capillary called a **glomerulus**. A long loop of the tubule leads out of the Bowman's capsule deep into the medulla, before returning to the

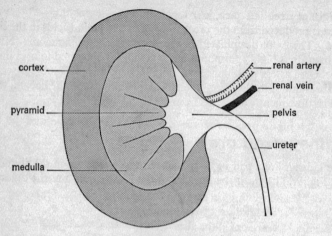

Fig. 9.2 Structure of the kidney

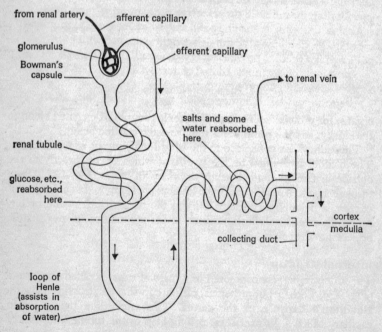

Fig. 9.3 Structure and function of a nephron

cortex. Many tubules then join together before emptying through a pyramid into the pelvis.

How the nephron works

1 Blood enters the glomerulus capillaries under high pressure, directly from the short renal artery. The blood pressure is also increased by (a) the narrowness of the coils in the glomerulus and (b) the difference in diameter between the afferent and efferent capillaries (Fig. 9.3).

2 **The high blood pressure** forces most of the smaller molecules carried in the plasma out of the glomerulus and into the thin-walled porous Bowman's capsule. These small molecules include water, glucose, amino acids, salts, urea, and other nitrogenous wastes. Red cells and large molecules such as proteins remain in the plasma. (**Note:** Blood does not flow through the nephron: the fluid entering the capsule is very different from blood.)

3 As the fluid filtered out of the glomerulus passes down the tubule, the cells lining the tubule reabsorb useful substances such as glucose and amino acids. These are passed back into the network of capillaries surrounding the tubule. This **selective reabsorption** takes place against a diffusion gradient (p55) and so requires expenditure of energy. The cells lining the tubule are richly supplied with mitochondria (p20) to provide energy. The ends of the cells projecting into the lumen of the tubule consist of microvilli (p19). This increases the surface area for absorption.

4 The remaining liquid is now composed mainly of urea, salts, and water. As it travels through the collecting duct, some of the sodium, salts, and water are reabsorbed, depending on the concentration of salts in the plasma at the time. If the plasma is unusually dilute (e.g. after drinking quantities of water) less water is reabsorbed, resulting in weak, pale urine. If the plasma is too concentrated, more water is reabsorbed, resulting in a strong, dark-coloured urine.

5 Selective reabsorption from the glomerular filtrate therefore removes all glucose and amino acids, together with some salts and water. The resulting liquid, now called **urine**, gradually trickles into the renal pelvis and is propelled by peristalsis to the bladder.

6 The capillaries in the medulla eventually unite to form the renal vein. The blood in the renal vein contains less water, salts, and urea compared to the blood in the renal artery, owing to their removal in the glomeruli.

The excreted substances in Table 9.3 are those whose concentration is higher in the urine than in the plasma, i.e. water, urea, and salts.

Table 9.3 Comparison of plasma and urine

Constituent	Contents in grams per 100 cm³ Plasma	Urine
Water	91·0	96·0
Urea	0·02	2·0
Various salts	0·42	1·18
Glucose	0·10	0·0
Proteins	6·5	0·0

Proteins and glucose are not normally excreted at all, because both are reabsorbed in the tubules.

Glucose, diabetes, and urine Glucose only appears in the urine if excess glucose is present in the plasma. This is a symptom of diabetes, caused by lack of the hormone insulin (p111).

Osmoregulation

Water is constantly lost from the body in urine, sweat, expired breath and faeces. This is regained by drinking, eating moist foods, and as a by-product of tissue respiration.

If too much water is lost (e.g. by sweating), or if excess salts are consumed, the osmotic pressure of the blood rises. The rise is detected by sensitive cells (**osmoreceptors**) in the hypothalamus of the brain (p106). These cause the posterior lobe of the pituitary gland (p110) to release **anti-diuretic hormone** (ADH) into the circulation.

The effect of ADH is to increase the permeability of the collecting duct to water, so that more water is reabsorbed and the urine becomes more concentrated. The release of ADH therefore reduces the loss of water in the urine and helps to maintain the osmotic pressure of the blood. If too much water is drunk, the blood's osmotic pressure falls and ADH secretion is reduced. This results in decreased permeability of the tubules, less reabsorption of water, and a more dilute urine.

Control of salt balance

Sodium ions are lost from the body fluids when sodium chloride is passed out of the body during sweating. If the plasma's sodium ion concentration falls, the hormone **aldosterone** is released by the cortex of the adrenal gland (p111). Aldosterone stimulates reabsorption of sodium ions in the tubules.

The bladder

The bladder is a muscular sac, whose walls contain elastic tissues allowing it to stretch to a volume of 400 cm^3. When this volume is reached, waves of contraction pass down the bladder and trigger a reflex, called the **micturition reflex**. As a result, impulses are sent via the spinal cord to the **sphincter muscle** at the mouth of the bladder (Fig. 9.1). In babies this causes the muscle to relax, so emptying the bladder. Older children acquire the ability to override the micturition reflex by means of voluntary commands from the brain, so that the bladder is emptied only when convenient.

Questions

1 *What is meant by excretion?* [p96; 2]
   List the various substances excreted by the body, explain the origin of each, and how each is excreted. [p.96; 10]
2 *Make a large labelled diagram of a kidney tubule, indicating the particular regions of the kidney in which the various parts of the tubule occur* [p98; 6]
   Describe concisely the functioning of the kidney in producing urine, referring particularly to the role of the kidney tubule. [7]
   What is osmoregulation? State briefly how the kidney acts as an osmoregulatory organ [5]   (CAM)
3 *Make a large labelled diagram of the urinary system, with the major blood vessels involved (internal details of the structure of the kidney are not required).* [p97; 8]
4 *The chief nitrogenous waste produced by man is urea, and not ammonia. Explain why.* [p.97; 2]
5 *Why do the cells of the kidney tubules contain a high concentration of mitochondria?* [p99; 2]
6 *Table 9.3 (p100) shows a simple comparison between some of the constituents of plasma and urine for a normal person.*
   (*i*) *Why does protein not appear in the urine?* [2]
   (*ii*) *Why does glucose not appear in the urine?* [2]
   (*iii*) *For which of the constituents listed in the Table would you expect to find a difference in a person whose pancreas produced insufficient insulin? Briefly indicate the differences which you would expect in the plasma and urine.* [2]

Answer guide
  (i) Protein molecules are too large [1] to pass through the capillary walls into the Bowman's capsule [1].

(ii) Glucose is reabsorbed [1] by the tubule cells as the urine passes through the nephron [1].

(iii) Additional glucose would be present in both the plasma [1] and the urine [1].

7   *What is the effect of warm weather on (a) the volume, (b) the concentration of the urine? Give reasons for your answer.* [p100; 5]

10   Co-ordination

The purpose of the nervous system is to co-ordinate the activities of the body, in response to stimuli both from inside and outside the body. The nervous system contains receptor cells (e.g. the retinal cells in the eye) sensitive to changes in the environment. Nerve cells transmit impulses from receptors to the brain and spinal cord, known together as the central nervous system (CNS); and from the CNS to effector organs. These are muscles or glands which respond to an impulse from a nerve.

The structure of nervous tissue

Nervous tissue is composed mainly of **neurones**. Each neurone (Fig. 10.1) consists of a **cell body**, which gives rise to numerous branching processes. Processes which conduct impulses away from the cell body are called **axons**; processes which conduct impulses towards the cell body are called **dendrites**.

Some neurones have an elongated axon, up to 1 metre long, or an elongated dendrite called a **dendron**. These elongated processes are also known as **nerve fibres**, and are usually surrounded by a fatty **myelin sheath**. This acts as an electrical insulator and speeds up the passage of the nervous impulse. The myelin sheath is composed of the cell membranes of **sheath cells** called **Schwann cells**.

There are three types of neurone: motor, sensory, and connector.

1 Motor (or efferent) neurones conduct impulses from the central nervous system to effector organs such as muscles and glands. Their cell bodies are situated in the central nervous system and the axon is greatly lengthened. Most axons terminate in **motor end plates**, each situated on a single muscle fibre.

2 Sensory (or afferent) neurones conduct impulses from the sense organs to the central nervous system. Their cell bodies are mainly

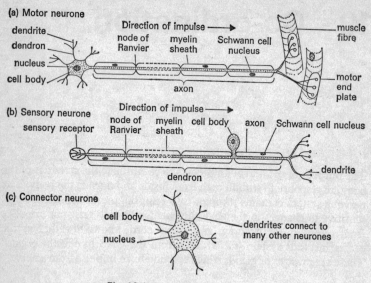

Fig. 10.1 Structure of nerve cells

located in a swelling, called a **ganglion**, situated at the junctions of the peripheral nerves and the spinal cord (*see* Fig. 10.4). The nerve fibre consists of a short axon carrying impulses into the spinal cord and a longer dendron carrying impulses from the sense organs to the spinal cord.

3 Connector neurones, also called relay, transfer, intermediate, or internuncial neurones, are located entirely inside the grey matter of the brain and spinal cord. They pass impulses from one to another, by means of a short axon and numerous branching dendrites.

Nerves are collections of nerve fibres (axons or dendrons) bound together with connective tissue. **Ganglia** (swellings), containing cell bodies, occur along the nerves.

The nerve impulse consists essentially of a wave of electrical activity passing along a nerve fibre. In a resting nerve, there is an electric potential of about 90 millivolts (mV) between the inside and the outside of the fibre. The inside carries a negative charge and the outside carries a positive charge. This potential is maintained by the outer membrane of the nerve fibre. As the impulse passes, the potential is temporarily reversed, so that a wave of altered electric potential travels along the fibre. (**Hint** Nerves do not 'carry messages'; they **conduct impulses**. Examiners will not accept the word 'message' in this context.)

Synapses are connections between nerve cells. The electrical impulse cannot cross the minute gap between two nerve cells, but when an impulse reaches a synapse a chemical is produced in the ends of the fibres. This diffuses rapidly across the gap between the cells and triggers off a fresh impulse in the cell on the other side.

The central nervous system

The central nervous system (the brain and spinal cord) is composed of grey and white matter. **Grey matter** contains mainly cell bodies and blood vessels, while **white matter** consists mainly of nerve fibres. The presence of the myelin sheaths makes the fibres appear white.

Cerebrospinal fluid
The delicate tissues of the brain and spinal cord are enclosed within tough protective membranes. These membranes secrete cerebrospinal fluid which circulates around the outside of the central nervous system, and also through a central canal. The fluid helps to protect the brain and cord by acting as a shock absorber.

Structure and function of the brain
The brain (Fig. 10.2) consists of the **fore-brain**, including the cerebrum, thalamus and hypothalamus; the **mid-brain**; and the **hindbrain**, including the cerebellum and medulla oblongata.

The **cerebrum** makes up the bulk of the human brain. It consists of two cerebral hemispheres, whose surface is greatly folded. The folded surface forms a complex outer layer called the **cerebral cortex**, which is richly supplied with blood vessels to provide food and oxygen.

The grey matter in the cortex consists of a layer surrounding the white matter. The deep folds of the cortex increase the surface area of the grey matter, so that it contains a large number of neurones. These are essential for the performance of complex mental tasks, e.g. reasoning and memory.

Functions of the cerebral cortex
These include:

1 **Interpretation** of sensory stimuli. For example, impulses from sensory cells in the eye are translated into a meaningful pattern.

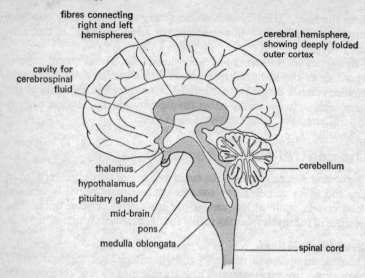

fibres connecting
right and left
hemispheres

cerebral hemisphere,
showing deeply folded
outer cortex

cavity for
cerebrospinal
fluid

thalamus

hypothalamus

pituitary gland

mid-brain

pons

medulla oblongata

cerebellum

spinal cord

Fig. 10.2 Section through the brain

2 **Association of stimuli** with memories of previous events, followed by **decision making** and **instruction** as impulses are passed out from the cortex to stimulate effector organs.

3 **Co-ordination** of the activities of the brain as a whole by means of connector neurones.

The sensory and motor functions of the cortex are known to be carried out in particular areas (*see* Fig. 10.3). The left cerebral hemisphere controls the right side of the body and vice versa.

The **thalamus** relays information from the lower centres of the brain, e.g. the cerebellum, to the cortex. It also controls the body's reaction to strong sensory stimuli such as pain, in the same way as the cerebral cortex. However, the thalamus is usually over-ridden by the cortex.

The **hypothalamus** is an area below the thalamus containing cells sensitive to the temperature, osmotic pressure (p96), and sugar concentration of the blood.

The **pituitary gland** (p110) is an outgrowth from the hypothalamus. It produces numerous hormones, usually following receipt of stimuli from the hypothalamus. The posterior lobe of the pituitary is directly connected to the hypothalamus by means of axons.

The **mid-brain** mainly contains fibres which connect the fore-brain to other parts of the brain and spinal cord.

The **cerebellum** is also divided into two hemispheres, again with the grey matter surrounding the white. Its function is to co-ordinate the actions of the skeletal muscles, so that complex movements (e.g. the maintenance of posture) are performed smoothly and efficiently.

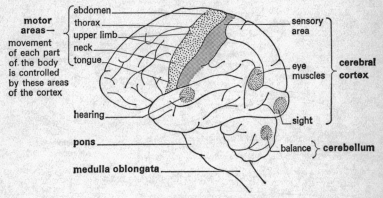

Fig. 10.3 Localized functions of the brain

The **pons** links the two halves of the cerebellum to each other and also connects the cerebellum to the central nervous system generally.

The **medulla oblongata** controls the automatic activities of the body such as digestion and heart beat. It also contains the respiratory centre (p55), which controls the breathing rate by means of cells sensitive to the carbon dioxide concentration of the blood.

The **spinal cord** extends from the brain almost to the sacrum (p36). It passes through the neural canal (p37) inside each vertebra. The cord contains grey matter in the centre (Fig. 10.4), surrounded by white matter. Pairs of peripheral nerves join the cord in the gaps between the vertebrae.

Response to stimuli

Reflex actions

A reflex action is a rapid, automatic, unlearned response to a stimulus. It is not controlled by the cerebral cortex. Reflexes are usually protective in nature, e.g. blinking to prevent damage to the eye, or coughing to remove unwanted material from the trachea.

How a spinal reflex works (Fig. 10.4) If the hand touches a hot object, pain receptors in the skin transmit impulses along sensory fibres to the spinal cord. The impulse is passed across a synapse to a connector neurone in the grey matter, and from there to a motor neurone. It then passes to motor end plates in the biceps muscle, causing muscle contraction.

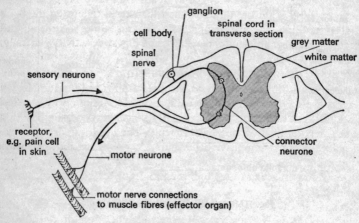

Fig. 10.4 Reflex arc

Voluntary actions

Voluntary actions involve conscious thought. They are controlled by the cerebral cortex in the brain and not by the spinal cord. For example, consider the case of swatting a fly. The movement of the fly is first detected by the retina of the eye. Cells in the retina transmit impulses via the optic nerve to the visual area of the cerebral cortex. These impulses are compared both to previous memories and to information from other receptors (e.g. the distance from the hand to the window). Impulses are then sent along connector neurones, via the cerebellum, to the spinal cord, and thence via motor neurones to the muscles.

Autonomic nervous system

The **autonomic nervous system** is a collection of nerves and ganglia concerned with the control of the unconscious activities of the body. It consists of two sets of nerves which connect the central nervous system to the internal organs.

The **sympathetic system** carries impulses which stimulate the organs concerned with immediate activity, e.g. running away. Thus the sympathetic system accelerates heartbeat, stimulates release of sugar from the liver, accelerates breathing, and diverts blood from the digestive organs to the muscles.

The **parasympathetic system** provides a parallel set of nerves to the same parts of the body as the sympathetic. However, parasympathetic impulses stimulate those organs which are active when the body is at rest. Thus the parasympathetic systems slows heartbeat and breathing rate, but accelerates peristalsis and digestive activity generally.

Effects of drugs on the nervous system

1 Depressants Drugs such as alcohol, aspirin, and tranquillizers depress the activity of the nervous system. Consumption of **alcohol** slows reactions to stimuli and particularly affects the cerebellum so that muscular co-ordination becomes difficult. The cerebral cortex is also affected, leading to impulsive behaviour. Alcohol causes dilation of the arterioles in the skin leading to loss of heat (p32). The effects may last for several hours, even to the following morning after heavy consumption at night.

Other depressants include aspirin, tranquillizers, barbiturates, and opium and its derivatives, morphine and heroin. Some of these drugs are medically useful because they help to relieve pain and anxiety through their depressant effect on nervous activity. However, some are dangerous drugs of addiction (e.g. heroin).

2 Stimulants increase the body's metabolic rate and are sometimes taken in preparation for a major event. Afterwards, the metabolic rate falls, leaving the individual feeling tired and depressed. Examples include caffeine, found in tea and coffee, 'pep pills', e.g. amphetamine and benzedrine, and hemp, with its derivatives cannabis and marijuana. Nicotine in tobacco smoke is also a stimulant.

The endocrine system

The endocrine system provides a means of co-ordination of body functions, parallel to the nervous system. In general, the endocrine system controls long-term body functions, e.g. growth, while the nervous system controls rapid, short-term activities, e.g. movement.

Endocrine (or ductless) glands are organs lacking ducts or tubes, so that their products (hormones) are passed directly into the bloodstream. Each **hormone** acts only upon particular organs or tissues, known as the **target organs**.

Endocrine glands should not be confused with exocrine glands. **Exocrine glands,** e.g. sweat glands, possess tubes (or ducts) leading out of the body or into the digestive system. Their products are secreted

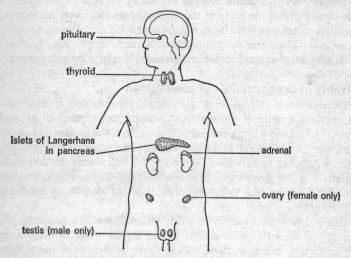

Fig. 10.5 Position of main endocrine glands

onto the surface of the skin or into the lumen of the gut, and not into the circulatory system. The **pituitary gland** produces numerous hormones, many of which affect the activities of the other glands – hence the description of the pituitary as 'the conductor of the orchestra' or 'the master gland'.

Important pituitary hormones These include:

1 **Thyroid stimulating (thyrotrophic) hormone, or TSH,** stimulates the thyroid gland to produce thyroxine.
2 **Follicle stimulating hormone (FSH)** stimulates the Graafian follicles (p.127) in the ovary to produce female hormone (oestrogen) and ova; in the male, FSH promotes production of male hormone (testosterone) and sperm by the testes.
3 **Adrenal cortex stimulating (adrenocorticotrophic) hormone, or ACTH,** stimulates the adrenal cortex.

4 **Luteinizing hormone** stimulates the ovaries and testes to release sex hormones; it also stimulates the ovary to produce a corpus luteum (p128).

5 **Pituitrin** (or growth hormone) stimulates growth, especially of the long bones. Most cases of stunted growth or excessive height are due to under- or over-production of pituitrin.

6 **Anti-diuretic hormone (ADH), or vasopressin,** increases the amount of water reabsorbed by the kidney tubules (p100).

7 **Prolactin** stimulates the mammary glands to produce milk.

8 **Oxytocin** stimulates both the release of milk and the contraction of the uterus muscles during birth.

The **thyroid gland** in the neck manufactures the hormonet **hyroxine**. This stimulates the metabolic rate, especially the rate of cell respiration. In children, it is essential for normal growth. Lack of thyroxine in childhood results in the victim growing up as a mentally retarded dwarf, called a cretin. Adults who lack sufficient thyroxine become slow and listless and feel cold. A common cause of thyroxine deficiencies is lack of **iodine** in the diet (p81), because iodine is an important constituent of thyroxine.

The **adrenal, or suprarenal, glands** lie just above the kidneys. The outer layer, or **cortex,** produces hormones such as **cortisone** which accelerates the conversion of proteins to glucose, among other functions. The inner part, or **medulla,** produces the hormone **adrenaline,** when stimulated by motor impulses direct from the brain.

Adrenaline has the same effect on the body as stimulation of the sympathetic nervous system (p109), i.e. the rates of heartbeat, breathing, and cell respiration all increase, blood is diverted from the digestive system to the muscles, and the pupils dilate. Adrenaline secretion is increased during moments of anxiety.

The **Islets of Langerhans** are small groups of endocrine cells scattered among the cells of the pancreas. The pancreas is therefore both an exocrine gland secreting digestive enzymes, and an endocrine gland secreting the hormone insulin.

Insulin reduces the level of glucose in the blood by accelerating its conversion to glycogen in the liver (p94). Insulin also promotes the uptake of glucose from the blood by the cells of the body generally. Lack of insulin leads to diabetes, in which the blood sugar level rises so that glucose is excreted in the urine. Urine is tested for glucose in all medical examinations.

The **testes** secrete **testosterone,** or male hormone, which stimulates the production of sperm and maintains the male secondary sexual characters (p130). The **ovaries** secrete **oestrogen**, or female hormone, which maintains the female secondary sexual characteristics and causes the lining of the uterus to thicken during menstruation (p130).

The **corpus luteum** (p128) in the ovaries also produces **progesterone** after ovulation. This promotes the growth of the wall of the uterus and prevents contractions of the uterus wall.

Integration of hormone action – carbohydrate metabolism

Many different hormones may affect various aspects of a particular process. Thus blood sugar level is raised by the action of cortisone stimulating the conversion of proteins to glucose, and by the action of adrenaline stimulating the conversion of glycogen to glucose. Blood sugar level falls through the action of insulin, stimulating the conversion of glucose to glycogen; and through the action of thyroxine, increasing basal metabolic rate, so accelerating the breakdown of glucose during cell respiration.

Questions

1 (a) *Draw labelled diagrams of sensory (afferent) and motor (efferent) neurones* [p104; 2 each]
  (b) *Describe one structural and one functional difference between sensory and motor neurones.* [4]

2 *Explain the differences between:*
  (i) *Stimulus and impulse;* [p103; 4]
  (ii) *Spinal cord and spinal column;* [p107; 4]
  (iii) *Grey matter and white matter;* [p105; 2]
  (iv) *Reflex and voluntary actions.* [p107; 4]

3 *Describe the main stages in a named reflex action – e.g. accidentally stepping on a nail.* [p108; 10]

4 *Name two reflex actions associated with the human eye, stating the stimulus and response in each case.* [p107; 7]   (WJ)

Answer guide
(a) Blinking [1]. *Stimulus:* rapid approach of an object to the eye [1]. *Response:* eyelid and facial muscles contract to close eyes [1].
(b) Contraction of pupil [1]. *Stimulus:* bright light [1]. *Response:* circular muscles [1] of iris contract to close pupil [1].
(**Note** The receptor involved in the iris reflex is the retina itself. There are no receptors in the iris.)

5 *Explain the difference between depressants and stimulants and give two examples of each.* [p109; 6]

6 *What are the general properties of hormones?* [8] See below.   (OX)

Answer guide

Hormones are chemicals [1] secreted [1] by endocrine glands [1], which circulate in the bloodstream [1]. They stimulate [1] the activity of a particular target organ [1]. Subsequently they are de-activated [1] in the liver [1].

7 *Explain the differences between exocrine and endocrine glands.* [p109; 4]

8 *In what ways do the products of named endocrine organs affect:*
  (*i*) *the growth and development of a child;* [p110; 5]
  (*ii*) *the ability of the adult to react to danger;* [p111; 5]
  (*iii*) *carbohydrate metabolism.* [p112; 5]   (JMB)

9 *Describe how body processes are effectively co-ordinated:*
  (*a*) *when an appetizing meal is placed before you;* [p87; 9]
  (*b*) *when you are suddenly frightened.* [p109; 6]   (WJ)

Answer guide

(a) The sight [1] and smell [1] of the food both set off reflex actions [1] which cause an increase in the flow of saliva [1]. In addition, the parasympathetic system [1] is stimulated. Nervous impulses transmitted along parasympathetic nerves [1] cause an increase in the flow of blood to the digestive system [1]; while the blood supply to other organs is reduced [1]; and peristalsis increases [1].

(b) Sudden fright causes stimulation of the sympathetic system [1], leading to increased rates of heartbeat [1]; breathing [1] and respiration [1]; and diversion of blood from the digestive organs to the muscles and brain [1]. Secretion of the hormone adrenaline also occurs and has the same effects [1].

11 Perception of stimuli: the senses

The body contains **receptors** (p29) sensitive to a wide variety of stimuli. This chapter describes the functions of receptors sensitive to light, sound, movement and orientation of the head, taste, and smell. Receptors sensitive to pressure, temperature, etc., are also found in the dermis (p27) and receptors sensitive to muscular tension ('stretch receptors') are located within both the voluntary and involuntary muscles.

The eye – structure and functions

Structure of the eye (Fig. 11.1)

The **eyelids** protect the eye by means of the blinking reflex (p112). This is induced by any sudden movement towards the eye.

The **tear glands** are situated above the eyeball and secrete tears (lachrymal fluid), a dilute salt solution which also contains the anti-bacterial enzyme lysozyme. The salt solution keeps the conjunctiva moist, so preventing drying and cracking, and washes away dust. Tears are secreted continuously and drain into the tear ducts, which open into the nasal cavity.

The **conjunctiva** is a thin epithelium which lines the eyelids and continues across the surface of the eyeball.

The **sclerotic** is a tough, non-elastic protective tissue enclosing the entire eye. It is mainly white, but it is transparent in the front (the cornea).

The **cornea** is the transparent part of the sclerotic layer. It is therefore protective in function. Much refraction of light occurs at the cornea (*see* image formation).

The **vitreous humour** is a thick, jelly-like fluid which fills the back of the eye. The pressure exerted by the jelly keeps the eyeball in shape.

The **aqueous humour** is a fluid occupying the front part of the eye.

It is more watery than the vitreous humour, but has the same function, since it maintains the shape of the front of the eye. In addition, it supplies food and oxygen to the cornea and lens.

The **lens** is a clear structure, held in place by non-elastic **suspensory**

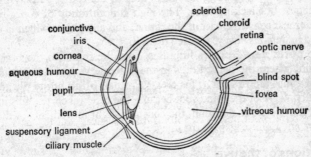

Fig. 11.1 Section through the eye

ligaments. These are attached to a ring of circular muscles called the ciliary muscles.

The **choroid** is a layer of tissue behind the retina containing a dark pigment which prevents the reflection of light within the eye. It also contains capillaries which provide food and oxygen for the cells of the retina.

The **iris** is an extension of the choroid. It consists of a pigmented disc containing two sets of involuntary muscles – the circular and radial muscles (Fig. 11·2). The function of the iris is to regulate the amount of light entering the eye.

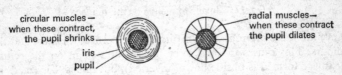

Fig. 11.2 Muscles in the iris

The **retina** contains light-sensitive cells of two kinds, rods and cones. **Rods** are extremely sensitive to light even at very low intensity. **Cones** are sensitive only to brighter light, but can discriminate between light at different wavelengths, i.e. different colours. Rods are more common towards the sides of the retina, so sensitivity to dim light is greatest at the sides of the eye.

Both rods and cones contain dark purple pigments which are bleached by light. This bleaching triggers off an impulse in the nerve cells connected to the retina. The pigments are regenerated immediately afterwards.

The **fovea** (yellow spot) is a small area in the retina densely packed with cones, but lacking rods. The high concentration of cones makes this the most sensitive part of the retina, and any object which is being studied in detail is always focused upon it.

The **optic nerve** carries impulses from the retina to the brain. At the point where the nerve fibres in the optic nerve leave the retina, no rods or cones are present. This creates a small **blind spot**. We are not normally aware of the blind spot because the cerebral cortex 'fills in' the gap in the image. In addition, the constant slight movements of the eye obscure any minute 'gap' in the image.

Functions of the eye
Image formation (Fig. 11·3)
When light moves from a less dense medium into a denser medium (e.g. from air to water) it is **refracted** ('bent') inwards. Light rays entering

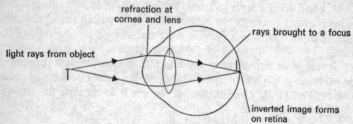

Fig. 11.3 Image formation on the retina

the eye are therefore refracted in turn by the cornea, the lens, and the humours. This causes the rays to be focused at a point on the retina. The image formed is upside down (inverted), but this is 'corrected' by the optic centre in the cerebral cortex (p105).

Accommodation (Fig. 11·4)
This is the ability of the eye to focus on objects at varying distances. It takes place as follows:

(a) *Adaptation for near vision*
1 The ciliary muscles contract inwards, overcoming the pressure exerted by the vitreous humour.

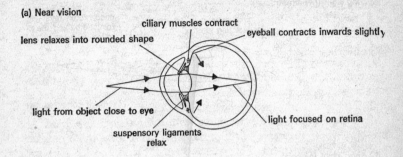

(a) Near vision

ciliary muscles contract

eyeball contracts inwards slightly

lens relaxes into rounded shape

light from object close to eye

light focused on retina

suspensory ligaments
relax

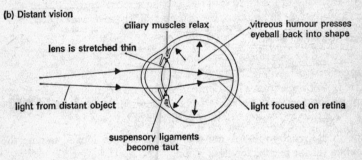

(b) Distant vision

ciliary muscles relax

vitreous humour presses
eyeball back into shape

lens is stretched thin

light from distant object

light focused on retina

suspensory ligaments
become taut

Fig. 11.4 Accommodation

2 As a result, the suspensory ligaments relax.
3 The lens also relaxes, and assumes its natural rounded shape. This shortens the focal length, i.e. it decreases the distance from the lens to the point where the image is in focus.

(b) *Adaptation for distant vision*
1 The ciliary muscles relax, and are pushed outwards by the pressure of the vitreous humour.
2 This pressure creates tension in the suspensory ligaments.
3 The tension in these non-elastic ligaments pulls the lens into an elongated, thin shape. This increases its focal length.

Defects of vision (Fig. 11·5)

Long sight (hypermetropia) In this condition, near objects cannot be focused. It is caused by a small eyeball, or a lens which has lost its elasticity. In either case, light from near objects is focused behind the retina. Loss of elasticity occurs especially in elderly people, when it is

(a) Long sight (hypermetropia)

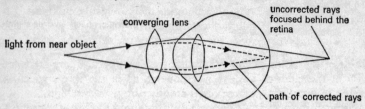

(b) Short sight (myopia)

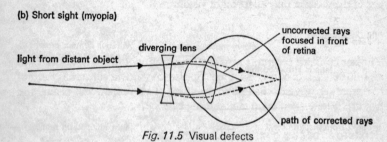

Fig. 11.5 Visual defects

known as **presbyopia**. A non-elastic lens can no longer assume the rounded shape needed for close vision.

Correction of long sight is achieved by wearing spectacles containing **converging** lenses, for close work, e.g. reading. These have the effect of decreasing the focal length of the eye as a whole.

Short sight (myopia) In this condition, distant objects cannot be focused. It is caused by too powerful a lens or too large an eyeball. This leads to light from distant objects being focused in front of the retina.

Correction of short sight is achieved by wearing spectacles containing **diverging** (concave) lenses. These increase the focal length of the eye.

Astigmatism With this condition it is impossible to focus horizontal and vertical lines at the same time. It is caused by irregularities in the curvature of the cornea and the lens, and is corrected by wearing cylindrical lenses.

Seeing in dim light
Sensitivity in poor light is increased by these means:

1 Dilation (widening) of the pupil, by contraction of the radial muscles of the iris. This lets more light into the retina and takes about 5 seconds to complete.

2 Adaptation to dim light in the retina After half an hour in dim light, the retina becomes 10 000 times more sensitive to low light intensity. This is probably due to a build-up of the purple pigment called rhodopsin in the rods. Rhodopsin is destroyed by light.

3 Use of the sides of the eye, rather than the fovea. There are more rods at the sides of the eye and none at all in the fovea, which is useless in dim light, because cones are insensitive to dim light.

Night blindness is a deficiency disease caused by lack of vitamin A (p78). Vitamin A is needed for the manufacture of rhodopsin, so shortage of the vitamin may affect night vision.

The ear

The ear contains receptors sensitive to sound vibrations of between 40 and 40 000 Hz depending on age. It also contains a separate group of receptors sensitive to changes in direction of movement and changes in posture. These are used in maintaining balance.

Structure of the ear

The ear (Fig. 11·6) is divided into three regions, the outer, middle, and inner ear. It is deeply set into the skull, for protection.

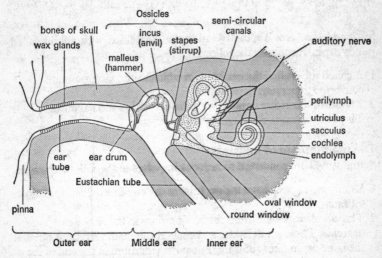

Fig. 11.6 Structure of the ear

(a) The **outer ear** contains the following parts:

1 A **tube** leading inwards (sometimes called the external auditory meatus). The walls of the tube contain wax glands which protect the ear. Small hairs lining the tube help to keep out insects and dust.

2 The **ear flap, or pinna,** attached to the outside of the tube, has little function in man.

3 The **ear drum, or tympanum,** is a thin membrane which stretches across the end of the tube. It is kept under tension by a muscle which pulls it inwards.

(b) The **middle ear** is an air-filled cavity in the skull, containing:

1 **Three small bones, or ossicles,** which connect the ear drum to a narrow opening called the oval window.

2 The **Eustachian tube** connects the middle ear to the nasal cavity. It is normally closed, but opens when swallowing or yawning. **Swallowing** allows air to leave the middle ear through the Eustachian tube so that the pressure becomes equal on both sides of the ear drum. If this does not occur, the ear drum becomes stretched taut under pressure either from the outside air or from the air within the middle ear. This prevents the drum from vibrating.

A bad cold can cause temporary deafness because excessive mucus may block the Eustachian tube. The 'popping' felt in the ears when swallowing during rapid pressure changes is due to the relief of pressure on the eardrum.

(c) The **inner ear** is a cavity inside the skull filled with a fluid called **perilymph**. It contains the following:

1 A coiled tube called the **cochlea,** in which sound vibrations are converted to nervous impulses.

2 The **semi-circular canals,** with the **utriculus** and **sacculus**. These are concerned with maintaining balance.

The mechanism of hearing

Sound waves are converted to nervous impulses in the following stages:

1 Sound waves pass down the ear tube to the **ear drum,** causing it to vibrate.

2 The vibrations of the drum set up vibrations in the **three ear ossicles.** These act as levers, and convert the small vibrations of the large drum (area: $85\,mm^2$) into large vibrations at the end of the tiny stapes (area: $3\,mm^2$).

3 **The stapes** vibrates against a thin membrane called the **oval window,** setting up vibrations in the perilymph. This sets up movements in the perilymph which pass down into the cochlea.
4 **The cochlea** consists of an inner tube called the cochlear duct, surrounded by two outer tubes. Vibrations from the oval window pass through the perilymph of the upper tube. They are then transmitted through a thin membrane into the fluid (endolymph) inside the cochlear duct (Fig. 11·7).

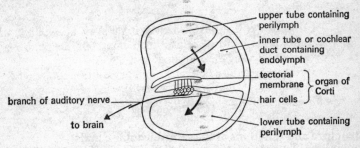

upper tube containing perilymph

inner tube or cochlear duct containing endolymph

tectorial membrane ⎫ organ of
hair cells   ⎬ Corti

branch of auditory nerve

to brain

lower tube containing perilymph

Fig. 11.7 Section through the cochlea. Large arrows indicate the path of vibrations caused by sound

5 The cochlear duct contains a layer of sensory cells. Each possesses a minute hair which is embedded in the **tectorial membrane** above them. This structure is called the **organ of Corti**. Vibrations in the tube above the cochlear duct set off vibrations in the tectorial membrane. The resulting tensions on the hairs trigger off the transmission of nervous impulses along the auditory nerve to the brain.
6 The vibrations then pass through the cochlear duct into the lower tube. They pass through the lower tube to the **round window** which vibrates into the air-filled middle ear.
The purpose of the lower tube and the round window is to dissipate the vibrations, i.e. to pass them harmlessly out of the enclosed, fluid-filled inner ear.

Sensitivity to pitch
High notes set off impulses from the sensory cells nearest the base of the cochlea, i.e. nearest the round window. Low notes are perceived nearer the tip of the cochlea. Impulses from high and low notes are transmitted to different parts of the cerebral cortex.

Balance

The inner ear contains the three semi-circular canals, which are sensitive to changes in the direction of movement, and the utriculus and sacculus, which are sensitive to changes in the position of the head.

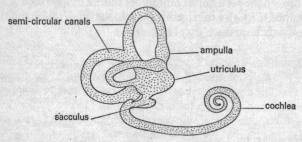

Fig. 11.8 The inner ear

Structure and functions of the semi-circular canals

The three canals are arranged at right-angles to each other. Each is filled with endolymph and each contains a swelling called an **ampulla** at the base. Inside the ampulla (Fig. 11·9) is a raised mound of sensory

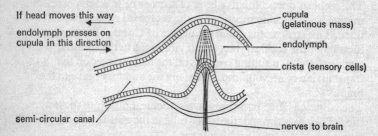

Fig. 11.9 Section through ampulla

cells called a **crista**. Each sensory cell ends in a sensory hair which is embedded in a gelatinous mass called a **cupula**.

Movement of the head produces slight pressure changes in the endolymph in the canals, so causing minute movements of the cupula. This sets up tension in some of the hairs, which triggers off impulses in the sensory cells.

Each canal is affected to a different extent by movement in particular directions. The pattern of impulses reaching the brain therefore provides accurate information about the direction of movements of the head.

The **utriculus and sacculus** also contain sensory cells with hairs embedded in a gelatinous plate, surrounded by endolymph. In addition, chalky granules called **otoliths** are present in the gelatinous mass. These weigh down the jelly, so pulling on the hairs when the head is tilted. The resulting nervous impulses provide information to the brain about the precise position of the head in space.

Information from the canals and from the utriculus and sacculus are combined with information from the eyes and the stretch receptors in the muscles to help maintain balance. Constant rotation (e.g. spinning round and round) may set up pressure in the canals which remains after motion has ceased. This results in the brain receiving conflicting information from the canals and the other senses and leads to feelings of giddiness.

Smell and taste

Olfactory (**smell**) **receptors** are found on the roof of the nasal cavity. They consist of sensory cells lying within the mucous membrane, with sensory cilia extending into the layer of mucus. These are **chemo-receptors**, sensitive to chemicals in the air which become dissolved in the mucus. Impulses from the cells are transmitted to the olfactory lobe of the cortex.

Much of the so-called 'taste' of food is actually due to the detection of smells given off by the food. Both taste and smell are much reduced during a heavy cold because of the extra layer of mucus covering the tips of the sensory cells.

Taste receptors are situated in the **taste buds** found along the sides of the papillae on the surface of the tongue. Each taste bud cell possesses a small protoplasmic strand which extends into the dissolved food and mucus mixture above it. Dry food cannot be tasted until it has been moistened with saliva.

Each taste bud is sensitive to one of only four sensations; sweet, sour, salt, and bitter. 'Sweet' and 'salt' buds are found at the tip of the tongue, 'bitter' at the back, and 'sour' at the sides. All other information about the taste of food is derived from the smell receptors.

Questions

1 *For each of the following, describe: (a) its location in the eye (b) its function;*

(i) *vitreous humour*, (ii) *choroid layer*, (iii) *cornea*, (iv) *rods*, (v) *cones*, (vi) *fovea*, (vii) *optic nerve*, (viii) *ciliary muscle*. (p115); 2 each]

2 *Explain in detail how the effect of shining a bright light into the eyes leads to a change in the size of the pupil.* [p115; 6]

Answer guide

Strong light shone onto the sensitive cells of the retina [1] causes transmission of impulses along sensory neurones [1] to the brain. This sets off a reflex action [1], in which impulses are transmitted via motor neurones [1] to the circular muscles [1] of the iris. These then contract, so reducing the size of the pupil [1].

3 *Describe the changes occurring in the retina after leaving a well-lit room and walking for 30 minutes in the darkness outside.* [p116; 2]

4 *Explain how the image of a stationary object is formed on the retina.* [p116; 6]

Answer guide

Do not forget that light is refracted by the cornea, and the aqueous and vitreous humours, as well as the lens. Use Fig. 11.3 as an illustration.

5 (a) *What is meant by 'visual accommodation'?* [p116; 2]

(b) *Explain how the eye accommodates for: (i) near vision; (ii) distant vision.* [p116; 5 each]

6 (a) *Draw a diagram to show the passage of light rays from a distant object through the eye of a short-sighted person.* [p118; 3]

(b) *Draw a diagram to show how this defect may be corrected by use of an appropriate lens.* [p118; 3]

7 *Describe with diagrams how sound waves reaching the outer ear are translated into nervous impulses transmitted along the auditory nerve.* [p120; 20]

Answer guide

Use the account given on p120, omitting the explanation for the purpose of the lower tube and the round window. Include Figs 11.6 and 11.7.

8 *How are we able to distinguish between sounds at different frequencies?* [p121; 2]

9 *Explain, with diagrams, how the following structures provide information about the movement of the head* [p122]:

(i) *semi-circular canals;* [8]

(ii) *utriculus and sacculus.* [8]

Explain how the body is able to distinguish between food with different tastes. [p123; 4]

12 Reproduction and child development

Structure and functions of the reproductive organs

The male
The male reproductive system (Fig. 12.1) includes the following organs:

1 The **scrotum** is a small sac containing the testes. It lies outside the abdomen so that the testes are about 2 °C cooler than the rest of the

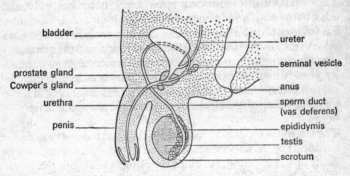

Fig. 12.1 Male reproductive organs (side view, in section)

body. This is essential for the development of the sperm. The **testes** produce male hormone or testosterone (p111) and sperm. The **epididymis** is a long (6 m) coiled tube on the outside of the testis. Sperm are stored here after manufacture.

2 **Sperm ducts** lead from the epididymis to a point below the bladder. Here they join together and unite with the tube leading from the bladder, to form the **urethra**. This is a tube leading from the bladder to the tip of the penis. It carries urine or sperm, though never both

together. (**Note:** Do not confuse the *urethra* with the *ureters*, which join the kidneys to the bladder.)

3 The **prostate gland**, the **seminal vesicles**, and **Cowper's gland** are three glands found along the sperm tubes and the urethra. As sperm pass through the tubes, these glands secrete fluids containing food substances and enzymes to activate the sperm. Until this occurs, the sperm are immobile. The resulting mixture of fluid and sperm is referred to as **seminal fluid** or **semen**.

4 The function of the **penis** is to transmit sperm from the male to the female. It consists of spongy connective tissue, richly supplied with blood vessels. During sexual intercourse, arterioles inside the penis dilate so that the spongy tissue fills with blood. This causes the penis to swell and become erect so that it fits more readily into the vagina. The tip of the penis contains numerous sensory cells which provide much of the sensation felt during intercourse.

The female

The female reproductive system (Fig. 12.2) contains the following:

1 The **ovaries** manufacture oestrogens (female hormones, p112) and other hormones, in addition to the ova (eggs).

2 The **oviducts or Fallopian tubes** end in funnel-shaped openings close to each ovary. Ova pass down the oviducts into the uterus.

3 The **uterus or womb** is normally a small structure about 80 mm long and 10 mm thick, containing involuntary muscle. It is capable of great expansion during pregnancy. The **cervix** (or neck of the uterus) is a ring of muscle which closes off the lower end of the uterus.

4 The **vagina** is a muscular tube which connects the uterus to the outside. It opens into a space called the **vulva,** which is enclosed by fleshy lips. A minute organ called the **clitoris** lies inside the vulva. It is richly supplied with sensory cells and provides sensory stimuli during intercourse.

5 The **urethra** carries urine from the bladder to the outside of the body. It opens into the vulva close to the vagina, but is otherwise entirely separate from the reproductive system.

The proximity of the urethra to the anus makes infection from microbes around the anal area more likely in females. **Cystitis,** a painful infection of the urethra and bladder, is often caused in this way. It is important to wash this area from 'front to back', i.e. from the urethra towards the anus, to prevent infection.

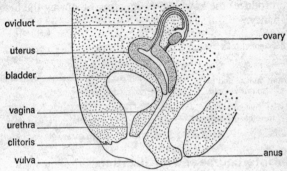

(a) Side view

oviduct

uterus

bladder

vagina

urethra

clitoris

vulva

ovary

anus

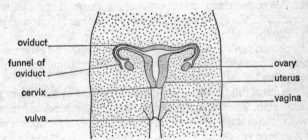

(b) From the front

oviduct

funnel of oviduct

cervix

vulva

ovary

uterus

vagina

Fig. 12.2 Female reproductive organs

Release of ova from the ovary—ovulation

From about the age of twelve onwards, the ovaries produce one ovum every month, alternately. When potential egg cells begin to mature in the ovary, the cells around them divide rapidly. A fluid-filled structure called a **Graafian follicle** forms (Fig. 12.3). This both nourishes the developing ovum and produces oestrogens (female hormones). Eventually the follicle projects from the surface of the ovary, bursts, and releases the ovum into the funnel of the oviduct. Ciliated cells lining the wall of the oviduct waft the ovum, together with attached follicle cells, into the tube and down towards the uterus.

Meanwhile, the follicle cells continue to grow, forming a solid mass of tissue called the **corpus luteum**. If fertilization occurs, the corpus luteum produces the hormone **progesterone** during the first 3 months of pregnancy.

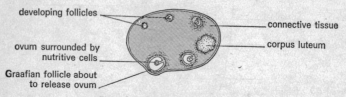

Fig. 12.3 Section through ovary

Sexual intercourse and fertilization

During sexual intercourse, or copulation, the following events occur:

1 Mutual sensory stimulation by both partners causes the male's penis to become **erect** by filling with blood. It can then be accommodated inside the vagina.

2 Contact between the tip of the penis and the walls of the vagina sets off reflex **muscular contractions** of the sperm ducts. This forces sperm out of the epididymis, past the prostate and the other glands, and into the urethra.

3 The prostate and adjacent glands add **nutrients and enzymes** to the sperm. These secretions stimulate the sperm to commence active swimming by lashing the tail.

4 The seminal fluid is **ejaculated** into the vagina by muscular contraction. The sperm then swim through the uterus and into the oviducts. Several million sperm are released at each ejaculation.

5 **If a sperm reaches an ovum,** the enzymes carried by the sperm disperse the follicle cells. The first sperm to reach the ovum also releases substances which dissolve the ovum's cell membrane sufficiently to permit entry.

6 The nuclei of the sperm and ovum then fuse to form a fertilized ovum or a **zygote**. After entry, a membrane develops around the fertilized ovum to prevent penetration by further sperm.

Fertilization takes place within 24 hours of copulation, since the ova die 24 hours after ovulation. For this reason also, it takes place in the part of the oviduct closest to the ovary.

Twins are formed when

(*a*) Two ova are present in the oviducts and both are fertilized (fraternal twins). **Fraternal twins** are no more alike than any brother or sister. They do not have identical genes and they do not share the same placenta.

(b) One ovum is fertilized by one sperm and later divides into two (identical twins). **Identical twins** are very similar because they have identical genes. They often share the same placenta.

Structure of the gametes (Fig. 12.4)

(a) Ovum

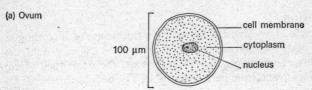

(b) Spermatozoon

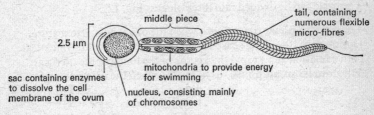

Fig. 12.4 Structure of the gametes

The **ovum** is a relatively large cell (over $\frac{1}{10}$th of a millimetre in diameter) containing food reserves in the cytoplasm. Each sperm is about forty times smaller than an ovum.

Puberty and adolescence

Between the ages of 10 and 16 in girls, and 12 and 18 in boys, the sex organs begin to produce the sex hormones. The testes and ovaries are themselves stimulated by hormones produced by the pituitary gland (p110). The sex hormones cause the development of the **secondary sexual characters**.

The **secondary sexual characters in girls** are:
1 Commencement of menstruation and ovulation.
2 Growth of breasts, uterus, and pelvis.
3 Growth of hair in the armpits and in the pubic region.
4 Development of feelings of attraction towards the opposite sex.

The **secondary sexual characters in boys** are:

1 Production of sperm.
2 Growth of muscles and penis.
3 Voice becomes lower in pitch.
4 Growth of hair in the pubic region, in the armpits, and on the face and chest.
5 Development of feelings of attraction towards the opposite sex.

Menstruation

A regular cycle of ovulation, followed by a thickening of the lining of the uterus, occurs on average every 28 days from puberty until the onset of the menopause (age 45–55). However, the cycle varies from woman to woman and from period to period. The purpose of menstruation is to prepare the lining of the uterus to receive the fertilized ovum. If fertilization does not occur, the lining is shed.

The cycle can be divided into three phases (Fig. 12.5):

1 The **growth phase** (days 6–14), when the ovum begins to mature. During days 10–14, the sharp rise in oestrogen secretion stimulates both ovulation and a rapid thickening of the lining of the uterus.
2 The **nourishment phase** (days 15–24), when the uterus lining

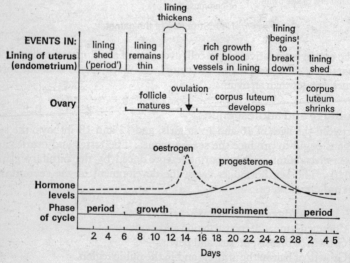

Fig. 12.5 The menstrual cycle

becomes richly supplied with blood vessels, under the influence of progesterone.

3 The **period** itself (days 1–5). The fall in progesterone level causes the shedding of the lining.

The menstrual cycle is a favourite topic for examiners, so study Fig. 12.5 carefully.

Pituitary control of menstruation (Fig. 12.6)

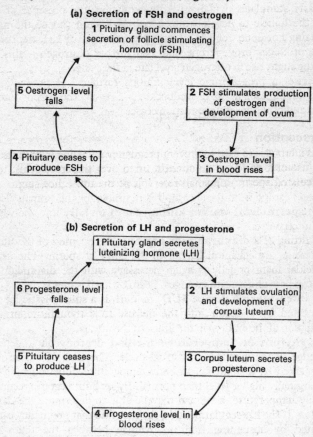

(a) Secretion of FSH and oestrogen

1 Pituitary gland commences secretion of follicle stimulating hormone (FSH)

2 FSH stimulates production of oestrogen and development of ovum

3 Oestrogen level in blood rises

4 Pituitary ceases to produce FSH

5 Oestrogen level falls

(b) Secretion of LH and progesterone

1 Pituitary gland secretes luteinizing hormone (LH)

2 LH stimulates ovulation and development of corpus luteum

3 Corpus luteum secretes progesterone

4 Progesterone level in blood rises

5 Pituitary ceases to produce LH

6 Progesterone level falls

Fig. 12.6 Hormonal control of menstruation. Note that a small rise in oestrogen level occurs late in the cycle, despite the absence of FSH. This is because the corpus luteum produces some oestrogen under the stimulus of LH.

During the first part of the cycle, the pituitary gland secretes **follicle stimulating hormone (FSH)** (p110). This stimulates the growth of an ovum in one of the follicles and also causes the ovary to produce oestrogen. However, the rising level of oestrogen inhibits further FSH production in the pituitary, eventually leading to a fall in the oestrogen level. This is called **negative feedback**.

The presence of oestrogen also stimulates the pituitary to produce **luteinizing hormone (LH)**. The combination of falling FSH and rising LH stimulates ovulation. After ovulation, the presence of LH causes the follicle to develop into a corpus luteum (hence the name 'luteinizing hormone' for LH). Under the influence of LH, the corpus luteum produces progesterone and also further oestrogen – this explains the slight rise in oestrogen level late in the cycle.

The rise in progesterone level inhibits further LH production – another example of negative feedback. As a result, the corpus luteum shrivels and progesterone production ceases.

Contraception

Various methods are used to prevent pregnancy following intercourse.

The use of a **condom or sheath**, fitted over the erect penis, traps the ejaculated sperm in a small reservoir at the tip. Since sperm may sometimes escape from the sheath, it is desirable for the woman also to insert a **spermicidal (sperm killing) foam or jelly** into the vagina before intercourse.

The fitting of a **diaphragm or cap** over the opening of the uterus also presents a mechanical barrier to the entry of sperm. The use of spermicidal foam or jelly is again necessary with the diaphragm, as sperm can pass underneath its edges. Foam by itself is ineffective.

The **intra-uterine device (IUD) or coil**, is a small plastic loop or coil inserted by a doctor into the uterus. It is thought to prevent implantation of he embryo in the lining of the uterus.

The **rhythm or temperature method** depends on a detailed knowledge of the woman's menstrual cycle. The time of ovulation is often marked by a slight fall in body temperature, followed by a rise to a marginally higher level than normal. By keeping a careful record of her daily temperature, a woman may be able to determine the day of ovulation. If she has regular periods, she may find that pregnancy can be prevented by abstaining from intercourse between the tenth and seventeenth days of the cycle.

In theory, the rest of the cycle can be regarded as 'the safe period' when the couple can have intercourse without taking precautions. In practice, the variations which occur in most womens' cycles make this an unreliable method.

The **contraceptive pill** is more reliable, since it prevents the release of an ovum. The pill contains chemicals similar to oestrogen and progesterone. If taken daily from days 5 to 25 in the cycle, these hormones inhibit ovulation. After the 25th day a period will follow owing to lack of progesterone.

Sterilization Men and women who have completed their families are increasingly requesting sterilization as a permanent means of preventing further pregnancies. In women, this is carried out by cutting and tying the oviducts, and in men by cutting the sperm ducts (vasectomy). Both methods are extremely effective, but cannot be easily reversed.

Effectiveness of contraceptive methods
The pill and the coil are highly effective. The use of the condom or the diaphragm, with foam in both cases, is also effective. The rhythm method is unreliable, as are the condom and diaphragm if used without foam.

Pregnancy

Early development Following fertilization, the zygote (fertilized ovum) floats down the oviduct, wafted along in the current created by

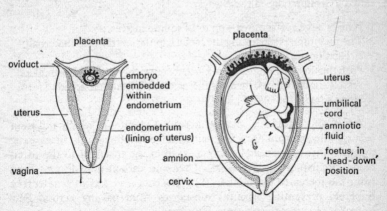

Fig. 12.7 Growth of the foetus in the uterus

the cilia lining the oviduct. During this time it divides rapidly to form a ball of cells. On entering the uterus, it floats freely for about 5 days, absorbing food and oxygen from fluids in the womb.

Implantation The embryo then becomes embedded in the wall of the uterus. Minute villi grow out from the embryo and become closely associated with the rich blood supply in the uterine lining (endometrium). Food and oxygen are absorbed through the villi from the maternal blood supply.

Later development The embryo grows rapidly so that the outlines of the limbs and major organs can be distinguished within 10 weeks of conception. By this time, the embryo is referred to as a foetus. After about 4 months, the foetus is able to move its limbs.

Function of the placenta

The minute villi formed during early development grow to form a large disc of tissue (the placenta), closely associated with the uterine lining and its rich blood supply. However, the maternal and foetal circulations always remain separate. If this were not the case, the higher blood pressure of the mother's blood would burst many of the delicate foetal capillaries. In addition, many substances in the mother's blood are poisonous to the foetus.

The **function of the placenta** is to secrete hormones, to absorb food substances and oxygen, and to pass foetal wastes, e.g. carbon dioxide and urea, into the maternal circulation. The structure of the placenta is suited to its functions in these ways:

1 It has a large surface area – up to 14 square metres.
2 The membranes separating the foetal capillaries from the maternal circulation are very thin.
3 Both the uterus lining and the placenta are richly supplied with blood vessels. In the uterus, many of these vessels become enlarged to form blood sinuses ('pools' of blood).

All of these features greatly increase the rate at which diffusion occurs. Glucose, amino acids, salts, and other food substances pass, together with oxygen and antibodies, from the maternal to the foetal blood circulation (*see* Fig. 12.8). Urea and carbon dioxide pass back from the foetus to the mother. The placenta also acts as a **selective barrier**, preventing harmful substances from passing across. Few microbes can cross the placenta, for example.

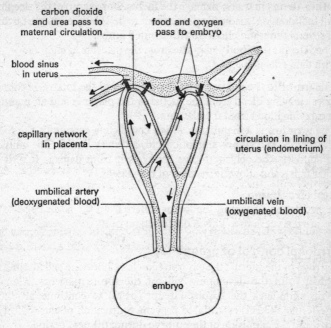

Fig. 12.8 Circulation in the uterus and placenta

Harmful substances affecting the foetus

However, some harmful substances and organisms can cross the placenta. These include:

1 The **rubella (German measles) virus,** which causes serious damage to the nervous system of the foetus if the mother catches rubella during the first 3 months of pregnancy. This may lead to the foetus being born blind, deaf, or with severe brain damage. Vaccination is essential before there is any risk of pregnancy.

2 **Smoking** during pregnancy also has a harmful effect, since it stunts the growth of the foetus. This leads to a smaller baby at birth, and so lowers the chances of survival if the baby is born prematurely. Evidently chemicals taken in from the smoke are able to cross the placenta.

3 A high level of **alcohol consumption** also has a damaging effect, since alcohol can easily cross the placenta. Babies with symptoms of alcohol poisoning are sometimes born to mothers who drink heavily.

4 **Other drugs** may also damage the foetus. For example, the sleeping pill thalidomide, medically prescribed in the 1960s, led to the birth of several thousand children with stunted limbs.

5 The **Rhesus antibody** may also cross the placenta, sometimes with fatal effects (*see* p64).

The **umbilical cord** connects the placenta to the foetus. It contains arteries carrying blood from the foetus to the placenta and an umbilical vein carrying blood from the placenta.

The **amnion** is a membrane which completely surrounds the foetus in the womb. It contains amniotic fluid, a watery liquid. The amniotic fluid supports the delicate foetus, protects it from damage (e.g. if the mother falls), and allows it freedom of movement.

Foetal circulation
The circulation system in the foetus differs from the adult because the lungs are not yet in use. This is described on p71.

Hormonal control of pregnancy
On arrival in the uterus, the embryo secretes a hormone called **gonadotrophin** from the developing villi of the future placenta. Gonadotrophin stimulates the corpus luteum (p128) to continue producing progesterone. This is essential to prevent contraction of the wall of the uterus, and the expulsion of the embryo during the next period.

The corpus luteum continues to produce progesterone as well as oestrogen, but after about 3 months, the placenta takes over this function. The presence of progesterone and oestrogen inhibit the secretion of FSH and LH (p132), so preventing further menstrual cycles during pregnancy.

Pregnancy tests
The first sign of pregnancy is therefore usually a missed period. To confirm this, a specimen of urine can be tested for gonadotrophin.

Needs of the pregnant woman
During pregnancy, special attention needs to be paid to these points:

1 **Diet** The diet needs to be rich in protein and minerals, especially calcium and phosphate for foetal bone growth and iron for the manufacture of red blood cells. Vitamins, especially vitamin C, are also needed to preserve both foetal and maternal health. 'Eating for two' is unnecessary and merely leads to obesity. A balanced, rather than a bulky diet, is needed.

2 Water Ample quantities of water should be drunk to stimulate kidney activity. This is important as the kidneys are responsible for removal of both maternal and foetal urea. Lack of water may lead to a rise in the level of urea in the mother's blood.

3 Clothing Clothes should be loose and comfortable so as not to interfere with the circulation. Regular mild **exercise** is desirable to stimulate the circulation and improve muscle tone.

4 Rest An afternoon rest is helpful to maintain peak health. The feet should be slightly raised to prevent the development of varicose veins.

5 Regular dental treatment during pregnancy is important, since infected teeth may produce poisons which affect the rest of the body. During pregnancy, dental treatment is free in Britain.

6 Alcohol should be taken only in moderation, and **smoking** avoided altogether.

Note: The section on ante-natal clinics, lactation, and child development may not be in your examination syllabus, so check carefully.

Ante-natal clinics

These provide advice and a programme of regular measurements to detect any abnormalities arising during pregnancy. These may include:

1 Toxaemia This is a condition affecting the kidneys. If left untreated, it leads to convulsions and premature labour. Symptoms include excess gain in weight, rise in blood pressure, and protein in the urine. Regular weighings, checks on blood pressure, and urine tests are necessary to detect early toxaemia. The treatment consists mainly of rest.

2 Diabetes (p111) The urine is also tested for glucose to detect early signs of diabetes. The excess glucose in the mother's blood may cross the placenta and so raise the osmotic pressure of the foetal blood. If untreated, this may lead to premature birth, or even the death of the foetus.

3 Anaemia (p58) Shortage of iron in the diet may cause anaemia in the mother and also in the foetus. Regular blood counts are therefore carried out to check on the number of red cells per cubic millimetre (mm^3) of blood. Iron tablets may be necessary to supplement the diet.

4 Blood group (p63) The mother's blood group is determined at an early stage, in case a blood transfusion becomes necessary. Emergency transfusions are sometimes needed owing to excessive bleeding at birth, and the blood group must be known in advance.

If the mother is Rhesus negative (p64), the father's blood is also tested. If he is Rhesus positive, severe complications may arise (*see* p65).

5 Chromosome defects, such as Down's syndrome (mongolism, p138) can be detected by inserting a hypodermic needle carefully into the uterus and withdrawing a few drops of amniotic fluid. By growing foetal cells from the fluid and examining them under the microscope, the chromosomes can be studied.

This test, called **amniocentesis,** is usually performed only on women who are thought to be at risk of producing a handicapped child. This includes women over 35 and those with a family history of Down's syndrome. There is no treatment for chromosome abnormalities, but the mother may wish to have the foetus aborted if it is likely to be severely handicapped.

6 Spina bifida Amniocentesis can also be used to detect this serious abnormality of the central nervous system because the amniotic fluid in a foetus affected by spina bifida contains certain unusual proteins. The diagnosis is then confirmed by use of an ultra-sound scanner. This uses sound waves to scan the foetus in much the same way as X-rays, but without risk of causing radiation damage to the foetus. No treatment is possible, but an abortion may be desirable.

The **size of the uterus** is also measured regularly. The information this provides on the exact stage of the pregnancy allows more accurate prediction of the expected date of birth. It also provides a guide to the general health of the foetus. Should the uterus suddenly cease to grow, this would indicate that the foetus may not be developing normally. Further tests would be needed to find out the cause.

Birth (parturition)

After about 9 months' gestation, signs of birth become evident. The following events take place:

(a) First stage

1 The foetus turns downwards so that its head lies just above the cervix.
2 The muscles of the uterus wall begin to contract rhythmically. The contractions become more frequent and more powerful. This marks the start of labour.

3 The opening of the cervix gradually expands, to admit the head.
4 The amnion breaks, releasing the amniotic fluid (the 'breaking of the waters').

(b) Second stage (birth)

1 Powerful contractions of the involuntary muscles of the uterus, aided by voluntary contractions of the abdomen muscles, force the foetus out of the uterus and through the vagina.
2 During birth, the foetus must pass through the pelvic girdle which forms a ring of bone. To aid this process, the separate bones of the foetal skull can slide over each other without harm.

In addition, the pelvis is wider in women than in men, and the cartilage joint at the front of the pelvis opens slightly (p38) during birth.

(c) Third stage

1 The umbilical arteries constrict, so that the cord can be cut without the baby bleeding to death.
2 Stimulated by the sudden change in its environment, the baby takes its first breath.
3 After 10 minutes, the placenta comes away from the wall of the uterus and is expelled as the afterbirth.
4 The circulation system of the baby begins to change to the adult arrangement (p71).

Lactation (production of milk)

Human babies, like those of all other mammals, must be fed on milk at birth. They lack teeth and their digestive system is unable to digest solids. Milk is manufactured by the **mammary glands, or breasts.** These are a pair of exocrine glands (p110), developed from the epidermis. They consist of milk-secreting cells, alveoli (small sacs) for the storage of milk, and ducts leading to the nipples.

Colostrum   During the first three days after birth, a thin lemon-coloured liquid called colostrum is produced by the breasts instead of milk. Colostrum is rich in protein but is not as nourishing as milk. It also contains antibodies (p61) which protect the infant from gut infections to some extent. At birth, the infant gut contains no bacteria, and the antibodies in colostrum prevent harmful types of bacteria from colonizing the ileum. During the first few months after birth, various harmless bacteria colonize the gut, where they assist in producing vitamin K (p78).

Composition of milk   Human milk is the best available complete food for growing babies. It contains most of the ingredients needed in a balanced diet for this stage of life, i.e. proteins, fat, carbohydrates, minerals, vitamins, and water, in the correct proportions. Iron is the only exception, but sufficient is usually present in the tissues of the new-born child for several months' growth (p81). A comparison of human and bovine (cow's) milk is given in Table 12.1.

12.1   Advantages of natural feeding

Natural feeding – human milk	Artificial feeding – cow's milk or preparations based upon it
Entirely suitable for digestion in the new-born human gut	Entirely suitable for calves but not ideal for humans
Human proteins are unlikely to cause allergies	Bovine (cow's) proteins are foreign to humans and may cause allergies
Contains most of the requirements of a balanced diet, in the correct proportions (except iron)	Mineral content is excessive, and may upset the balance of salts in the body
Contains correct proportion of protein	Excessive protein content reduces acidity in stomach, so increasing risk of gut infections
Contains correct quantity of lipases	Must be sterilized, which destroys lipases needed for digestion of fat and absorption of the fat soluble vitamins A and D
Contains human antibodies which give some protection against infection during early life	Contains bovine antibodies, which provide little protection against human diseases
Requires no preparation or sterilization	Inconvenient to prepare and sterilize
Breast feeding helps to develop a close relationship between mother and child	Bottle feeding cannot provide so close a relationship
Free	Must be purchased

The **disadvantages of breast feeding** are:

1   The father cannot assist with feeding. He therefore cannot share the night feeds, so that the mother may become over-tired. There is also less opportunity to build up a father–child relationship.
2   Many women find breast feeding difficult owing to cracked or infected nipples. Some babies also seem to prefer to feed from a bottle.

Conclusions Breast feeding is clearly superior to bottle feeding; but it is not practicable for every mother. However, even a few days' breast feeding is helpful since this will allow for the transmission of antibodies.

Child development

Diet in childhood

Weaning is the term given to the gradual withdrawal of breast or bottled milk and its replacement with solids. Solids can be added after about 4 months. This also helps to reduce the number of feeds.

Any solids given to a child between 4 and 6 months old must be entirely free from lumps. Babies cannot chew before 6 or 7 months because the chewing muscles are not yet developed. After 6 months, they can chew using their firm, hard gums – even though they have no teeth. They can be given any suitable foods such as biscuits, cheese, bananas, etc., provided these do not contain bones or other hard objects.

Balanced diet for children

The basic rules for the choice of a child's diet are similar to the requirements of a healthy adult's diet (p80). It is desirable to continue providing milk, since this contains most of the requirements of a balanced diet; but all bottles and teats used must be carefully sterilized. After the first few months, cow's milk is less likely to cause any unexpected illness.

Children require much less food than adults (p82). A one-year-old may need only 4000 kilojoules daily, compared with 11 000 for an average adult. Overfeeding may lead to obesity, so excess carbohydrates should be avoided.

Specialized diets

A small number of children are born with genetic abnormalities for which special diets are needed. Children with phenylketonuria are born unable to break down the amino acid phenylalanine, owing to the presence of a faulty liver enzyme. This may lead to brain damage caused by excessive phenylalanine and its by-products.

The blood of all 1-2-week-old babies is therefore tested for the presence of phenylalanine by-products (and also for diabetes, p111). If by-products are present, the child must be fed on a diet containing minimal phenylalanine until the age of 5. By then, the brain is no longer affected.

Sleep

A new-born baby sleeps throughout the 24 hours, except when being fed. This gradually reduces to 14 hours at night, with an hour's rest in the morning and afternoon, at 1 year. By the age of 4, the morning rest will have long ceased, and the afternoon rest becomes less regular. Up to the age of 10, about 12 hours sleep per night is needed, slowly reducing to the adult average of 7½ hours.

Growth in height and weight

Growth occurs unevenly, in spurts, and varies greatly from person to person. It also differs in boys and girls, especially between 10 and 16. The chief milestones in growth are as follows:

1 **A fall in weight** of up to 10% occurs during the 2 days immediately after birth. Rapid growth soon makes up for this within a week or two.
2 **The maximum rate of growth** occurs during the first year of life (Fig. 12.10). Between birth and 12 months old, a typical baby

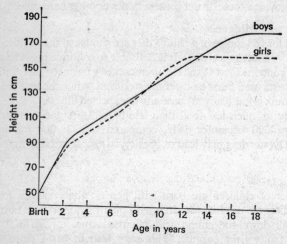

Fig. 12.9 Growth in height for boys and girls

increases in height by at least 25 cm, and in weight by 6 kg.
3 **The rate of growth** (*see* Fig. 12.10) then falls rapidly until puberty. At this point, there is a growth spurt in girls between the ages of 11 and 14, and in boys between 13 and 16.
4 **Boys are slightly taller** than girls on average, up to the age of 11 (Fig. 12.9). Between 11 and 13, girls are taller on average, but boys catch up and are finally 12 cm taller on average.

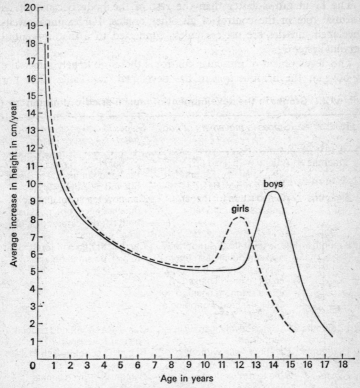

Fig. 12.10 Growth rates in boys and girls

5 **Girls complete growth** in height by about 16 on average, but boys usually grow taller until about 18 years of age.

6 **Increase in weight** may continue in both sexes as the shoulders broaden and the muscles develop.

Growth rates of different organs

Different parts of the body grow at different rates. The head of a new-born baby is very large in proportion to the rest of the body, owing to the rapid growth of the brain. This continues in childhood so that by the age of 4, the brain has reached 90% of its adult weight, while the body as a whole is only about a quarter of its adult weight. Later the other parts of the body grow faster than the brain until the adult proportions are achieved.

The brain grows faster than the rest of the body because of its essential role in the control of all other organs. For example, well-developed muscles are useless unless connected to a mature central nervous system.

The development of muscular control is therefore largely a measure of how far the nervous system has developed (*see* Table 12.1) For

Table 12.1 Stages in the development of muscular co-ordination

Age	Stages in movement of body	Stages in development of hand-eye co-ordination
By 3 months	Able to hold head up; able to sit up, if supported, for short periods	Can hold a toy if placed in the hand but unable to reach out and pick up toys
By 6 months	Able to sit up for longer periods, if supported	Can now reach out and pick up a large toy, but must use both hands; cannot pick up small objects
By 9 months	Able to sit up unsupported and can pull herself into a sitting position; begins to crawl; can pull herself into a standing position and can stand if gripping a support	Can use thumb and forefinger to pick up small objects
By 12 months	Crawls actively; can walk if supported	Can use a spoon and so feed herself (clumsily)
By 18 months	Can now walk without support	Begins to scribble with a pencil
By 2 years	Able to run	Can dress if given simple clothes
By 2½ years	Can jump with both feet and walk on tiptoe	Can undress completely, except for buttons; can use manipulative tools, e.g. scissors
By 3 years	Can stand on one foot	Can dress and undress almost completely

example, when a baby first tries to pick up a toy it often misjudges the distance. Eventually, the child learns by experience how to co-ordinate the sensory information coming from the eyes with the motor impulses needed to produce the correct muscle movements.

Development of speech

From the age of 8 weeks, babies begin to make noises. At about 9 months, they may begin to imitate the sounds made by their parents, e.g. 'dadda', 'mamma' but these will not necessarily have any meaning yet.

By 12 months, a baby usually understands the meaning of words and many can say two or three words (e.g. 'mama' or 'teddy'). By 18 months, a child can understand sufficiently to go on simple errands, e.g. fetching a toy. Many children of this age can also say numerous single words. By the age of 2, the child can put two or three words together to make a sentence – a major step in the development of speech.

Bladder and bowel control

Young children initially have no voluntary control over the sphincter muscles which close the bladder (p101) and rectum. The development of voluntary control usually occurs at about 18 months. Even then, the child can only delay urination for a few minutes after he has felt the desire. By 2 years old, most children can give reasonable warning and fetch the pot themselves; by $2\frac{1}{2}$ years, they can use the toilet for both urination and defaecation.

As with all stages in development, there is much variation in the rate of toilet training. Control over defaecation occurs before control over urination, and girls learn both earlier than boys, on average.

Questions

1 *With the help of large, labelled diagrams, describe the path of a sperm from its production in the testis to the time it fertilizes an egg in the female body.* [p126; 6, 6, 7]   (OX)

Answer guide
It is not immediately clear how these marks are to be allocated, but it is probably 6 each for the diagrams of the male and female systems, and 7 for the description. Use diagrams such as Figs 12.1 and 12.2(b). No description of any organ need be given, since this is already included in the diagram.

2 (a) *Give an illustrated account of the changes which take place during the menstrual cycle* (p130) *in:*
      (i) *the endometrium (lining of the uterus),* [5]
      (ii) *the ovary.* [7]   (CAM)
   (b) *Name the hormones that influence the menstrual cycle and briefly describe their functions.* [4]
   (c) *Name the substances contained in a given type of contraceptive pill and explain the mode of action of these substances.* [4]
3 *What is the role of the pituitary gland in the control of menstruation?* [p131; 8]

Answer guide
With only 8 marks available, it is essential to be brief. For example:

During the first part of the cycle, the pituitary gland secretes follicle stimulating hormone (FSH) [1], which stimulates the growth of an ovum [1] in the ovary [1] and also promotes production of oestrogen [1] by the ovary. In the second half of the cycle, the pituitary produces luteinizing hormone [1], which stimulates ovulation [1] and the formation of a corpus luteum [1]. The corpus luteum then produces progesterone [1].

4　(a)　*How is the foetus supplied with oxygen and food?* [p134; 10]
　(b)　*How are carbon dioxide and urea removed from the foetal circulation?* [4]

5　*Give two reasons why it is desirable for the placenta to act as a selective barrier between the maternal and foetal circulations.* [p134; 4]

Answer guide
It is essential to prevent harmful substances passing into the foetal circulation. These include pathogens [1] and maternal blood [1]. The maternal blood pressure is too strong for the foetal blood vessels [1], and mingling of the two circulations might cause an antigenic reaction which would destroy the foetal red cells [1].

6　*Describe with diagrams:*
　(a)　*the route taken by the fertilized ovum from the place where conception usually occurs until implantation has taken place.* [p133; 8]
　(b)　*the development of the foetus in the uterus up to, but not including, birth.* [17]

7　*What advice should be given to a woman concerning her health during pregnancy? Give a reason for each suggestion.* [p136; 10]

8　*Why is regular attendance at ante-natal clinics important for maternal and foetal health?* [p137; 2] *List five measurements or tests carried out at ante-natal clinics and explain the purpose of each.* [10]

9　*What is the role of the placenta and the pituitary gland in the onset and maintenance of lactation?* [p111; 8]

10　*Figure 12.10 on p143 shows the rate at which the height of average boys and girls increases annually with age.*
　(i)　*At which ages do* (a) *girls,* (b) *boys, grow fastest?* [2]
　(ii)　*At which age do both sexes grow least?*
　(iii)　*Comment on the differences in rates of growth between boys and girls at different ages, as shown in Fig. 12.10.* [3]

Note: Remember that the graph shows rates of growth and not actual height. In answering questions about graphs or tables of information, base your answer only on the information given in the graph. Do not answer from your own general knowledge.

11 *Outline the main stages in development of muscular co-ordination and speech from birth until the age of 3.* [p144; 10]

13   Inheritance

Cell division

Cell division is the means by which growth of the body occurs; in growing tissues each cell grows to a certain size and then divides into two. In addition, the genetic material is passed from cell to cell during this process.

Cell division occurs in all kinds of cell during the growth of a foetus and the early years of childhood. Later, as cells become specialized they may lose the power of cell division, e.g. nerve cells. In a few tissues cell division continues throughout life, e.g. the germinative layer in the skin (p28) and the bone marrow cells which manufacture red blood cells.

Mitosis

All cells, except gamete mother cells (p156), divide by means of mitosis. This results in two new cells, each with an equal share of chromosomes and other cell organelles.

Mitosis has five stages, as shown in Fig. 13.1. (**Note:** Many examination boards do not require the names of the stages to be known. For these syllabuses mitosis can be described as one continuous process.)

1 Interphase

This is the resting phase between divisions. It lasts between 10 and 24 hours in rapidly growing tissues. During this stage there is no visible sign of division, but the following events take place:

(a) The chromosomes multiply, so doubling the quantity of DNA in the cell. This provides sufficient DNA for the two daughter cells.
(b) New cell organelles such as mitochondria and centrioles (p19) are manufactured.

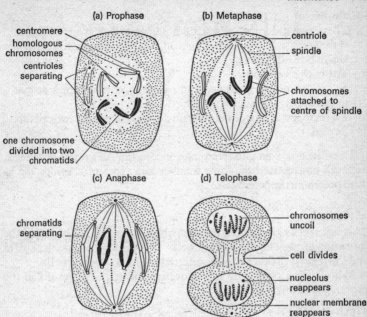

Fig. 13.1 Stages in mitosis

2 Prophase

(a) The chromosomes shorten and thicken, so becoming visible under the light microscope. Each chromosome now consists of a pair of identical structures called **chromatids,** joined together by a single centromere.

(b) The nuclear membrane and the nucleoli disappear.

(c) The two centrioles separate and move to opposite ends of the cell.

(d) A fibrous structure called a **spindle** begins to appear between the chromosomes.

3 Metaphase

(a) Spindle formation is completed.

(b) The chromosomes become arranged along the equator (middle) of the spindle, still in the form of paired chromatids.

4 Anaphase

(a) The paired chromatids begin to separate.

(b) One partner from each pair gradually moves to the opposite end of the spindle.

5 Telophase

(a) The chromatids, now called chromosomes, collect at opposite ends of the spindle. They gradually uncoil, become thinner, and can no longer be easily seen.

(b) The nucleoli reappear.

(c) Nuclear membranes form around each set of chromosomes, so that there are now two nuclei present.

(d) The cytoplasm between the two nuclei constricts until two cells are formed.

The cells now pass into interphase once again. As a result of mitosis, each cell has received the same number and types of chromosome as were present in the parent cell.

Meiosis

This is a special kind of cell division which takes place only during the formation of gametes, i.e. the **sex cells**. In humans these are the sperm and ova (eggs); all other cells are referred to as the **somatic cells**.

Meiosis takes place only in the ovaries and testes. It results in the production of cells which have:

1 half the number of chromosomes found in the parent cell;
2 different combinations of chromosomes from those in the parent cell.

The **reduction in chromosome number** is necessary because the fertilized ovum contains the chromosomes from both the male and female gametes. Human cells normally contain 46 chromosomes; if the gametes also contained 46, then the zygote (fertilized egg) would contain 92.

The **different combinations of chromosomes arise** because the chromosomes become reshuffled during meiosis. This can produce individuals who do not resemble their parents closely.

The stages in meiosis (Fig. 13.2)

Meiosis involves two consecutive divisions. By the end of the **first division** (Fig. 13.2a), there are only half the total number of chromosomes at each end of the spindle (if we ignore the chromatids at this stage). The first division is therefore called the **reduction division**, since the number of chromosomes is reduced from the original number (46), known as the **diploid number**, to the **haploid number** (23). The haploid number is always half the diploid number.

During the **second meiotic division** the chromosomes divide again

(a) First meiotic division

(i) Pairs of homologous chromosomes lie close together. Each chromosome is already composed of a pair of chromatids.

(ii) Homologous chromosomes separate except at the chiasmata. Chromatids exchange material by breaking and reforming at the chiasmata (crossing over)

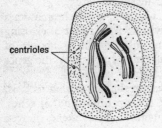

centrioles

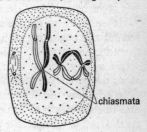

chiasmata

(iii) Nuclear membrane disappears. Spindle forms.

(iv) Chromosomes migrate to ends of spindle, taking exchanged material with them.

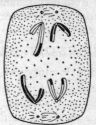

(b) Second meiotic division

(i) The chromosomes divide again at right angles to the first division. This causes the chromatids to separate.

(ii) Four new cells are formed, each containing the haploid number of chromosomes.

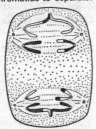

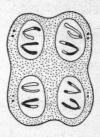

Fig. 13.2 Stages in meiosis

at right angles to the first division, but this time the chromatids are separated.

The second meiotic division is similar to mitosis, though with two separate divisions inside one cell. Four new cells are formed, each with one complete set of chromosomes. These are derived from the chromatids at the beginning of the second meiotic division.

However, these chromosomes are not identical to the chromosomes seen at the beginning of first meiotic division. As a result of crossing over at the chiasmata, each chromosome may contain a mixture of sections derived from both parent chromosomes. In this way, the genes along the chromosomes have been separated and recombined.

In addition, the chromosomes have been reshuffled during meiosis, and this also leads to fresh combinations (**recombinations**) of genes. At the start of meiosis, the parent cell contains 23 chromosomes derived from its male parent and 23 derived from its female parent.

The cells formed at the end of meiosis (gametes) contain only 23 chromosomes each. These are a random assortment of chromosomes derived from each parent. For example, all 23 might come from one parent, or 12 from one and 11 from the other, etc.

The laws of heredity

The genetic material

The basic unit of inheritance is the **gene**. A gene is a small part of a chromosome, made mostly of DNA (p23). Each gene carries the instructions for making a particular protein. This, in turn, determines an inherited feature such as blood group or eye colour.

Most genes exist in a number of different forms called **alleles** (or **allelomorphs**). For example, the genes which produce brown eye colour and blue eye colour are both alleles. Each allele is carried on the same place on homologous chromosomes. Since homologous chromosomes are paired, there are normally two alleles of the same gene present in each person.

Thus most people are born with two alleles for eye colour, in any of these combinations:

B = the allele which produces brown eyes
b = the allele which produces blue eyes

BB	**Bb**	**bb**
Two brown eye alleles	Mixed alleles	Two blue eye alleles

People with mixed alleles (**Bb**) usually have brown eyes, because the allele for brown eyes masks the allele for blue eyes. An allele which masks the action of another allele is called a **dominant allele**. The weaker allele is referred to as a **recessive allele**.

If a person bears two identical alleles for the same character, e.g. **BB**, they are said to be **homozygous** for that character. If a person bears two different alleles for the same character, e.g. **Bb**, they are said to be **heterozygous** for that character.

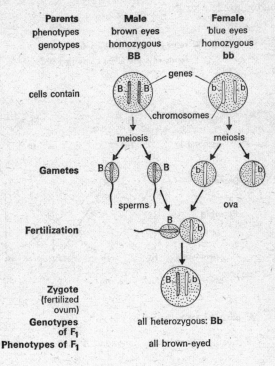

Note: **F₁** means first generation after the parents.

Fig. 13.3 Inheritance of eye colour

Phenotype and genotype

The observed features of a person are referred to as their **phenotype**; their genetic constitution is referred to as their **genotype**. People with brown eyes all have the same phenotype; but their genotypes may be different, since some are homozygous for eye colour (**BB**) and some are heterozygous (**Bb**).

How inheritance works

Up to 85% of white people carry a protein called Rhesus factor in their red blood cells (Rh⁺) (p64). The other 15% do not possess this factor and are called Rhesus negative (Rh⁻). The allele for Rh⁺ is dominant to the allele for Rh⁻.

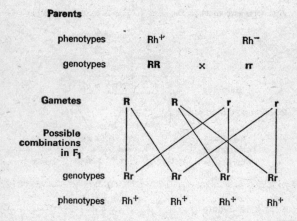

R represents the allele conferring Rh⁺
r represents the allele conferring Rh⁻

Parents

phenotypes   Rh⁺   Rh⁻

genotypes   **RR**   ×   **rr**

Gametes   R   R   r   r

Possible combinations in F₁

genotypes   Rr   Rr   Rr   Rr

phenotypes   Rh⁺   Rh⁺   Rh⁺   Rh⁺

All offspring are heterozygous and Rh⁺

Fig. 13.4 Results of a marriage between a homozygous Rh⁺ and a homozygous Rh⁻

The results of a marriage between a homozygous Rhesus positive man and a Rhesus negative woman are shown in Fig. 13.4. Figure 13.5 shows the possible results of marriages between individuals who are heterozygous for this factor.

When the children from large numbers of marriages between heterozygous couples are studied, they show a ratio of three Rh⁺ to one Rh⁻.

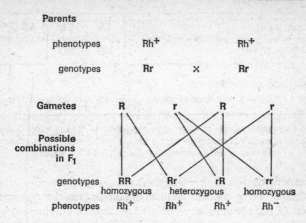

Parents

phenotypes   Rh⁺   Rh⁺

genotypes   Rr   ×   Rr

Gametes   R   r   R   r

Possible
combinations
in F₁

genotypes   RR   Rr   rR   rr
           homozygous   heterozygous   homozygous

phenotypes   Rh⁺   Rh⁺   Rh⁺   Rh⁻

A simpler way to show the combinations is to use a table like this:

Gametes	R	r
R	RR	Rr
r	rR	rr

Possible combinations in F₁

genotypes   RR   Rr   rR   rr

phenotypes   Rh⁺   Rh⁺   Rh⁺   Rh⁻

Fig. 13.5 Possible results of a marriage of two heterozygous Rh⁺

This 3:1 ratio only applies if the offspring from large numbers of families are considered; it may not apply to individual families.

Incomplete dominance

Some alleles are not completely dominant to each other. This occurs with the ABO blood group system (p63). Everyone belongs to one of four major blood groups: group A, group B, group AB, or group O. Three alleles are responsible, called A, B, and O.

The A and B alleles each produce a particular protein on the red blood cells. In the AB blood group both proteins are produced, and the two alleles are therefore 'co-dominant'.

The O allele produces no characteristic protein and is recessive to A and B. This results in the following combinations:

Combinations of alleles in the genotype of the children	AA or AO	BB or BO	AB	OO
Resulting blood groups appearing in their phenotype	A	B	AB	O

Fig. 13.6 Possible results of a marriage between an AO man and BO woman. If large numbers of marriages of this kind are studied, equal numbers of children result in each group (i.e. a 1 : 1 : 1 : 1 ratio).

Sex determination

Every human being contains 22 pairs of homologous chromosomes and one pair of sex chromosomes. In women, the sex chromosomes are a pair of homologous chromosomes called X chromosomes. Men have only one X chromosome, paired with a smaller Y chromosome.

At meiosis, each gamete receives 22 'ordinary' chromosomes (auto-

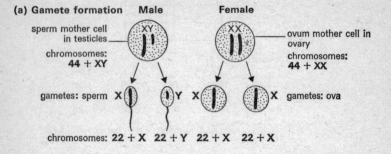

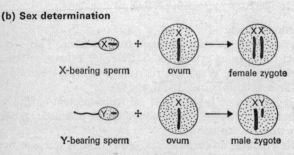

Fig. 13.7 Gamete formation and sex determination

somes) plus one sex chromosome. Half of the male gametes (sperm cells) receive an X chromosome, the other half receive a Y chromosome. All female gametes (ova) receive one X chromosome each (Fig. 13.7a).

The sex of a zygote depends upon whether it is fertilized by an X-bearing sperm or a Y-bearing sperm. Sex is therefore determined entirely by the male gamete. Roughly equal numbers of boys and girls are produced because equal numbers of X-bearing and Y-bearing sperm are produced in the testes (Fig. 13.7b).

Sex-linked genes

Since men carry only one X chromosome, any recessive gene inherited on the X chromosome cannot be masked by a dominant allele on the homologous chromosome. This means that men can inherit a number of abnormalities which are rarely found in women.

For example, about 8% of British men suffer from red–green colour blindness, because the gene is carried only on the X chromosome. If a woman inherits the allele for colour blindness, it is usually masked by the allele for normal vision. When this occurs, she is referred to as a **carrier**; i.e. a person carrying an allele for a particular condition but not showing it herself.

B represents the allele conferring normal vision

b represents the allele conferring colour blindness

— indicates the absence of an allele, owing to the lack of the second X chromosome in men

Parents

| phenotypes | normal man | carrier woman |
| genotypes | B— | × | Bb |

Gametes	B	b
B	BB	Bb
—	B—	b—

Possible combinations in F$_1$

genotypes	BB	Bb	B—	b—
phenotypes	normal woman	carrier woman	normal man	colour blind man

Fig. 13.8 Inheritance of red–green colour blindness

If large numbers of children from marriages of the type shown in Fig. 13.8 are studied, all the girls have normal vision, but half are carriers for the colour blindness gene. Half the boys are normal, and half are colour blind. Colour-blind girls can occur only from the marriage of a colour-blind man to a carrier woman.

Other sex-linked genes include that for **haemophilia,** a disease in which the correct proteins for rapid blood clotting are not produced. This leads to serious bleeding even after a trivial bruise. The haemophilia gene is usually passed on by female carriers; on average, half of their sons will be haemophiliacs.

Mutations

A mutation is a spontaneous change in a gene or a chromosome. This often results in alterations in the phenotype. For example, haemophilia is caused by a mutation in a gene which controls the production of a protein required for effective clotting of the blood. Phenylketonuria (p141) is caused by a mutant gene which contains an incorrect code for the manufacture of an essential liver enzyme.

Mutations are caused by various types of radiation, e.g. X-rays, ultraviolet light, cosmic rays, atomic radiation, and also by exposure to certain chemicals. These may damage the DNA (p23) in the chromosomes.

Mutations can occur in any cell at any time, but they are not usually important unless they happen during gametogenesis (i.e. the production of sperms and ova), or during early cell divisions in the newly fertilized zygote.

Mutant genes produced at these times may be transmitted to many cells in a growing foetus. Unnecessary exposure to radiation should be avoided during early pregnancy; and the ovaries and testicles should not be exposed to radiation in adults of child-bearing age. However, mutations are rare: only about one in a hundred thousand genes mutate, on average.

Mutations affecting chromosomes (chromosome aberrations)

Mutations occasionally affect whole chromosomes or pieces of chromosome. For example, a mutation may cause the chromosomes to divide unevenly during meiosis. This could result in an ovum with one extra chromosome, i.e. 47 instead of 46.

The presence of this extra chromosome may lead to serious abnormalities in the child, including a weak heart, an easily infected chest, and severe mental retardation. This type of handicap is known as

Down's syndrome, or **'mongolism'** owing to the distinctive shape of the eyelids.

Questions

1 *What are homologous chromosomes?* [p152; 2]

2 *What happens to: (a) the nucleolus, (b) the nuclear membranes, (c) the chromosomes, during the initial stage of mitosis?* [p149; 4]

3 *Where does (a) mitosis, (b) meiosis, occur, and why is each necessary?* [p150; 6]

4 (a) *Briefly explain the meaning of the following terms:*
[p152; 2 each]
heterozygote   dominant   phenotype
homozygote   recessive   genotype
mutation   allele

(b) *List the sex chromosomes found in (i) males, (ii) females.* [p156; 2]

(c) *Explain why approximately equal numbers of males and females are born at birth, with reference to the laws of genetics.* [p157; 6]

5 *Explain how it is possible for two brown-eyed parents to produce a child with blue eyes.* [p153; 7]

When answering genetics problems

1 You should use symbols such as **R** or **r** for the alleles, but you must always explain what each symbol stands for.

2 You should set out the symbols as in Fig. 13.5 so that you show every possible combination of gametes. A table is easier to draw than a diagram using arrows.

6 (a) *If curly hair is dominant to straight hair, what proportions of curly- and straight-haired children would you expect in a family where the father was heterozygous for curly hair and the mother had straight hair?* [p153; 1]

(b) *Show on a diagram how you calculated your answer, using the symbols:*
H for the gene for curly hair and h for the gene for straight hair.
[p153; 4]   (ox)

7 *Explain, by using suitable symbols, how parents with normal colour vision can have a son who is red–green colour blind, and other sons and daughters who have normal vision.*
(Red–green colour blindness is a sex-linked condition.) [p158; 8]
(JMB)

14 Prevention of disease

This chapter deals with infectious disease, its causes and prevention; and with the National Health and Local Authority services available for the sick and the vulnerable.

Causes of infectious disease

Human infections are mainly caused by various kinds of micro-organisms. The main categories involved are:

1 **Viruses,** very small organisms which can only live inside other living things. They exist by invading other cells and causing the host cells to manufacture virus protein.
2 **Bacteria** are larger than viruses and are mainly free-living saprophytes, although some are the cause of serious diseases. They multiply rapidly by dividing in two. They contain DNA scattered throughout the cell, so they do not possess a nucleus.
3 **Protozoa,** e.g. Amoeba, are single-celled animals, mainly found in water, where they feed on bacteria or microscopic plants. Only a few cause human illness; malaria is the most famous Protozoan disease.
4 **Fungi** are mostly free-living saprophytes, but a few cause human skin diseases.

Most microbes are either harmless or beneficial to man. Those that cause disease are known as **pathogens**. (**Note:** The word 'germs' should not be used in exams.)

A few other, larger, organisms also cause human diseases, e.g. external and internal parasites such as fleas and tapeworms. Some syllabuses require a knowledge of these, but there is much variation between the exam boards.

Spread of pathogenic organisms

Owing to their small size, microbes spread easily by the following means:

1 Droplet infection During exhalation (breathing out), a fine spray of moisture is passed into the air. This may carry pathogens, especially viruses, from an infected person. Coughing, sneezing, spitting, and shouting are especially likely to spread infected droplets up to a metre away.

2 In dust The spores of bacteria and fungi may be blown about in dust and so enter the respiratory system. Regular removal of dust, both at home and at work, helps prevent this. Breathing through the nose (p49) also helps to filter dust from the air.

3 Insects Many kinds of microbe are spread by flies (p167). **Fleas,** carried by rats, may carry the bacteria which cause bubonic plague. The **body louse** (p169) may carry the serious disease typhus, caused by organisms similar to bacteria. **Mosquitoes** carry the malaria parasite (p164) and the virus which causes yellow fever.

4 Other animals Pets and farm animals may carry pathogens. Close contact with infected animals may lead to illness.

5 Faulty personal hygiene Microbes can easily pass onto the hands following defaecation, so it is essential to wash the hands after using the toilet and before preparing food.

6 Faulty domestic hygiene Kitchens need special attention because of the risk to food (*see also* p179). Draining boards should be made from smooth, non-absorbent surfaces, e.g. metal; but preferably not wood. Wood absorbs water containing microbes, and therefore needs to be regularly disinfected. Pitted surfaces and chipped and cracked crockery trap bacteria and food. Sinks must be kept free of small pieces of food. Washing up should always be carried out in the hottest possible water, with added detergent. All working surfaces in kitchens need to be regularly washed down with bleach or other disinfectants to kill microbes.

7 Waste matter **Faeces** must always be disposed of in a hygienic manner (p186). **Dry refuse** (p187) should be placed in clean, dry dustbins with secure lids. This prevents entry of flies, and, by excluding moisture, also helps to prevent bacteria from breeding.

8 Drinking water Many microbes spread easily in water, especially if it is contaminated with faeces. Purification of water supplies (p184) is essential.

9 Food Moist food, especially if kept warm, provides ideal conditions for bacteria to multiply. This often leads to food poisoning (p178).

Prevention and cure of infectious disease

Defence against disease
The outside of the body is protected in the following ways:

1 The skin is protected by secretions of **sebum** which are mildly antiseptic, and by the presence of the hard, dry, epidermis.
2 The ears are protected by **wax** and the eyes by **tears,** which are also antiseptic (p162).
3 The respiratory and reproductive passages are protected by secretions of **mucus.**
4 The gut is protected by the secretion of **acid** in the stomach.

Treatment of disease
Serious infections can be treated by injection of antibodies against the invading microbe concerned. **Antibiotics** (e.g. penicillin) may also be given. These are substances originally found in fungi which prevent bacteria (but not viruses) from growing.

Antibiotics should not be confused with **antiseptics**. Antiseptics are chemicals used to kill microbes; but they may also be harmful to human tissues. Mild antiseptics can be used on the skin, or as a component of cough sweets, etc. Strong antiseptics (e.g. disinfectants) can kill all living cells and so should not be applied to open wounds, or consumed, unless in a very dilute form.

Public health and the prevention of disease
Various terms are used to describe the state of a particular disease in a particular country:

1 An **endemic disease** is a disease which occurs normally in a particular country, giving rise to a steady flow of cases every year. The common cold is endemic to Britain.
2 An **epidemic** occurs when the number of cases of a disease suddenly rises above normal.
3 A **pandemic** occurs when an epidemic suddenly sweeps across several countries and the number of cases rises to levels even higher

than in most epidemics, e.g. bubonic plague in the fourteenth century.

To prevent epidemics, various public health measures are employed:

1 Notification of diseases Doctors are required by law to report cases of certain dangerous 'notifiable' diseases, e.g. typhoid and diphtheria, to the health authorities. This allows immediate action to be taken when the first case occurs.

2 Isolation Anyone infected with the disease, or merely suspected of infection, must be kept away from other people, e.g. by removal to an isolation hospital.

3 Quarantine Isolation must be continued for as long as the individual remains infectious, even though the symptoms of illness have long disappeared. Anyone who has been in contact with the disease ('contacts') must also be isolated. To determine the length of quarantine for contacts, it is necessary to know the **incubation period** for the disease. This is the length of time which elapses between the entry of the pathogen and the onset of the first symptoms (e.g. 14–21 days in chicken pox). Contacts must be isolated for the duration of the incubation period, starting from the last date of contact.

4 Carriers Carriers must be traced, as well as people with symptoms. A carrier is a person who is infected with a disease but does not display symptoms. Carriers may infect other people, often for many years, without any sign that they themselves are infected.

5 Source of infection This must also be traced. Sterile 'swabs' (e.g. cotton-wool mounted on a stick) are rubbed over possible sources of pathogens (e.g. food, faeces, surfaces in a house, etc.) and then rolled carefully over sterile agar plates. Any microbes present will grow on the agar and so can be identified.

6 Mass vaccination (p62) This can also be organized to prevent further cases arising.

Survey of infectious diseases

This section describes the infections most frequently mentioned in exam questions.

Athlete's foot is a fungus disease affecting the skin, especially areas likely to be permanently moist, e.g. between the toes. It causes red, irritable areas and may spread widely if not treated. Its spores spread rapidly from person to person in wet conditions. It is therefore likely to spread in changing-rooms, especially after baths or showers. The sharing of towels is particularly likely to hasten its spread.

Preventive measures include regular washing of changing-room floors with disinfectant, use of chlorine in swimming baths, and careful drying of the feet between the toes. Treatment with antibiotic creams is usually effective, after a time.

The common cold is caused by a virus which attacks the respiratory passages, leading to a slight fever (rise in temperature) and increased mucus production. Secondary infection by bacteria may follow, leading to catarrh or sinus infections.

The virus spreads rapidly in droplets, and is especially likely to penetrate the body's defences during sudden spells of cold weather, or when a person is 'run down' generally. There is no medical cure, and immunity is short lived as there are many different varieties of cold virus. Use of disposable tissues is important to prevent its spread.

Diphtheria is caused by a bacterium which does not penetrate the body deeply, but releases powerful toxins (poisons). It attacks the throat, and may hinder breathing owing to the formation of a membrane over the tonsils. The toxins may also cause general poisoning of the body, leading to death.

Diphtheria is especially dangerous in young children. It spreads in droplets or by contact with infected bedding. Treatment includes the injection of diphtheria anti-toxin to neutralize the toxins. Vaccination (p62) has greatly reduced the number of cases in Britain since 1940.

Malaria is caused by a protozoan (p160) called *Plasmodium*, and is transmitted by female *Anopheles* mosquitoes. The parasite is found especially in the salivary glands of the mosquito. When the mosquito bites, it injects saliva containing an anti-coagulant to prevent the blood from clotting, so facilitating feeding.

At each bite hundreds of parasites may be injected into the blood with the saliva. They multiply in the liver, and later in the red blood cells. When released into the bloodstream, they cause periodic fevers. These may continue for many years, causing great weakness, and even death if combined with the effects of starvation or other diseases.

Treatment is by means of specific anti-malarial drugs to kill the parasite. Prevention is achieved by keeping mosquitoes away from people and preventing them from breeding. Mosquitoes fly mostly at night and lay eggs in still water; the larvae and pupae develop in water, before becoming adult.

The following **preventive methods** are used:

1 Taking **anti-malarial drugs** in malarial areas as a matter of routine, to kill any parasites as soon as they enter the bloodstream.
2 Fixing **mosquito nets** over doors and windows at night.
3 Spraying houses with **insecticides,** e.g. DDT, to kill the adult mosquitoes.
4 **Spraying** ponds and standing water with oil to prevent the larvae from breathing at the surface.
5 **Draining** casual water (e.g. marshes, gutters, old buckets, etc.) to prevent breeding.

Malaria is mainly a tropical disease because a warm climate is necessary for rapid multiplication of both the mosquitoes and the parasite.

Pneumonia is the name given to a variety of lung infections, mostly caused by bacteria. It leads to a high fever with inflammation of the lung. The bacteria involved are often found living normally in the respiratory passages; they usually only become infectious in those with low resistance, e.g. the very young, the very old, or individuals recovering from a severe illness. Treatment with antibiotics aids recovery.

Ringworm is a skin disease caused by fungi, most often in children between 5 and 10. It is especially common on the scalp, where it may spread in a ring-like pattern. The fungus grows in the hair shafts, causing the hair to weaken and fall out. Scratching may also lead to secondary infections. Ringworm spreads by close contact with infected cattle, dogs, cats, and people. It can be treated by means of a specific antibiotic.

Rubella (or German measles) is a mild virus disease which causes only a slight rash and mild fever in adults. However, it may cause serious harm to the early foetus (p135). Vaccination for girls aged between 11 and 13 is therefore essential.

Sexually transmitted infections (venereal diseases) affect the genital organs and are passed on during close sexual contact. **Gonorrhea** is caused by a bacterium which invades the urethra. This results in inflammation leading to pain on passing urine, especially in men. In women, the organism is more likely to invade the vagina, and the disease is often initially symptomless. If untreated, it may lead to sterility due to blockage of the oviducts. **Non-specific urethritis (NSU)** is similar to gonorrhea and equally common, but its cause is unknown.

Syphilis is a more serious disease caused by organisms similar to bacteria, known as spirochaetes. The first symptom is the appearance of a sore called a chancre. Later a rash appears, and after some years, the disease may cause severe damage to the nervous system and other organs.

All three major sexually transmitted infections can be cured with antibiotics. Since the organisms concerned can only survive in warm, moist conditions, these infections spread only from close personal contact. They can easily be prevented by restricting sexual relations to uninfected partners. Should infection or contact with an infected person occur, it is essential to seek medical treatment.

Tetanus is caused by a bacterium normally found in the gut of horses and sheep, so that its spores are common on cultivated land. The tetanus bacterium is anaerobic and therefore grows especially in deep, dirty wounds. Its powerful toxins interfere with the working of the nervous system, causing sudden muscular spasms. These frequently lead to death – 100 people are killed by tetanus in Britain every year. Treatment consists mainly of injections of anti-toxin.

Anyone taken to hospital with a dirty wound is automatically vaccinated against tetanus. However, even a small, unnoticed puncture wound may lead to infection and death, so it is important for all children to be vaccinated and for adults to have occasional booster doses.

Tuberculosis (TB or consumption) is caused by a bacterium. It most commonly infects the lungs, slowly destroying them and leading to death if untreated. Symptoms include general weakness, coughing up blood, and a characteristic appearance of the lungs on X-ray plates. The disease spreads in droplets produced by coughing and spitting.

TB is associated with poor living standards, especially malnutrition. It is therefore becoming rare in developed countries, though still endemic in Britain. Treatment with antibiotics, combined with rest in pleasant surroundings, is usually effective.

Typhoid fever is caused by a bacterium which infects the intestines. It causes a high fever with severe diarrhoea. The bacterium spreads easily through water supplies contaminated by faeces; it is also spread by flies or even by individuals who handle food without previously washing their hands. It is often spread by carriers (people who are carrying the bacteria without exhibiting symptoms).

Typhoid is nowadays unknown in developed countries, except when brought in from abroad. Its decline is largely due to the separation of drinking water from water carrying faeces, together with the treatment of both drinking and waste water (p184). Vaccination is essential for anyone visiting countries with ineffective water treatment facilities.

Other organisms affecting man

Certain exam syllabuses require a knowledge of the life cycles of some of the species listed below – but check your syllabus carefully first.

Insects	Flatworms	Roundworms (Nematodes)
Housefly	Tapeworm	*Ascaris*
Flea	Blood fluke	*Oxyuris* or *Enterobius*
Louse	Liver fluke	(pin or thread worm)

The housefly
Houseflies spread a variety of pathogens, including the bacteria which cause typhoid, cholera, and various gastric diseases (e.g. 'summer

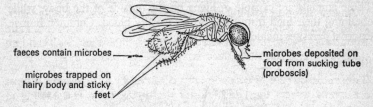

faeces contain microbes

microbes deposited on food from sucking tube (proboscis)

microbes trapped on hairy body and sticky feet

Fig. 14.1 How flies spread disease

diarrhoea'). Flies easily become contaminated with microbes because they feed on all kinds of organic material, including faeces, human food, and rotting wastes.

They feed by secreting digestive enzymes onto their food and then sucking it up. They also deposit spots of vomit or their own faeces on human food; in addition, microbes become easily trapped on their hairy bodies and sticky feet (Fig. 14.1). Flies are therefore capable of depositing microbes wherever they walk, including plates, cups, etc.

Life-cycle of the housefly The eggs are laid in small groups in any warm, damp, organic material, and hatch after about 8 hours to produce the white limbless larvae known as maggots. These burrow into the organic material, and feed on it. After moulting twice, the larva develops into a pupa, from which the adult hatches after 3 days.

Houseflies are rare in winter because in cold temperatures their life cycle is completed very slowly. Since microbes also multiply slowly at low temperature, gastric infections spread by flies are commoner in summer – hence 'summer diarrhoea'.

Control of houseflies This is best achieved by the prevention of breeding, i.e. by denying them access to sources of organic matter near houses. Methods of control therefore include:

1 Manure and compost heaps should be kept as far as possible from houses.
2 Dustbins should be kept covered, and sprayed with insecticide in warm weather.
3 Organic refuse should be thoroughly buried.
4 All food should be kept covered.
5 Insecticidal sprays can be used against the adults, but with caution, as these may be harmful to people and animals.

External parasites (ectoparasites)
These animals are capable of living on the outside of the body, while obtaining their food from their host – usually blood. Common human external parasites include fleas and lice.

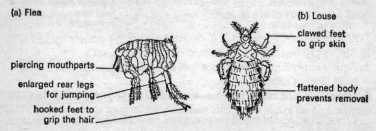

Fig. 14.2 Human ectoparasites

The **human flea** (Fig. 14.2(a) is a wingless insect, with its back legs greatly enlarged for jumping. The feet are hooked to enable the flea to grip human hair. Fleas feed on blood, using their piercing, needle-like mouthparts to penetrate the epidermis.

Their bites cause irritation, leading to itching with the possibility of secondary infection of the wound by bacteria. Fleas can be prevented by good personal hygiene and controlled by use of insecticidal powder.

Lice (Fig. 14.2(b) are also wingless insects. They grip the body closely with their clawed feet and feed on blood, using their piercing and sucking mouthparts. The victim's scratching may lead to serious skin diseases, e.g. impetigo (an infection of the skin by bacteria). In addition, lice spread typhus and other diseases.

There are three kinds of louse found in Britain:

1 **The head louse** lives in the hair and lays eggs called nits, stuck to the hair. They hatch out in about 6 days as nymphs, which are essentially miniature adults. After a further 10 days, the nymphs moult, become adult, and breed. Contrary to popular belief, the head louse is just as common in clean, short hair as in dirty, long hair, and spreads easily whenever people are in close contact (e.g. in schools). Treatment is by use of insecticidal shampoos and combing with a special comb to remove the nits.
2 **The body louse** is found only among those who rarely bath or change their clothes (e.g. tramps). It lays eggs on the clothes and is only found on the skin when feeding.
3 **The pubic or crab louse** is usually found in the pubic hair, and is often spread during sexual intercourse. Both the body and the pubic louse are otherwise similar to the head louse.

Internal parasites (endoparasites)
These are parasites which live inside the body. Their special adaptations for this way of life include:

1 Ability to resist digestion (e.g. worms living in the gut).
2 Ability to resist the host's antibodies (blood parasites).
3 Ability to produce large numbers of eggs, since transmission to new hosts is often difficult.
4 Use of hooks or suckers to cling to the wall of the gut (e.g. tapeworms).

In addition, sexual reproduction poses problems for internal parasites since they cannot leave the host to search for a mate. Tapeworms are therefore self-fertilizing, while male and female blood flukes remain coupled together permanently once they have come together inside the host.

Tapeworms (e.g. *Taenia solium* – the pork tapeworm) These live in the human intestines, attached to the gut wall by means of hooks and suckers. They have no digestive system, but feed by absorbing digested food. The body (Fig. 14.3) consists of numerous segments in which reproduction takes place by self-fertilization. The segments (proglottids) fill with eggs, become detached, and pass out with the faeces.

If eaten by a pig, the eggs hatch into larvae which bore through the gut and form thin-walled sacs called **bladderworms** in the muscles. They remain dormant until eaten by man, when the bladderworm develops into the adult tapeworm. The symptoms of infection include feelings of nausea, loss of weight and general weakness due to lack of food.

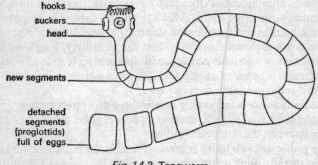

hooks
suckers
head
new segments
detached
segments
(proglottids)
full of eggs

Fig. 14.3 Tapeworm

Control of tapeworms involves:

1 visual inspection of killed meat by environmental health officers (p188) – the bladderworms are visible to the eye;
2 thorough cooking of all meat;
3 high standards of personal and public hygiene to ensure proper disposal of faeces.

Blood flukes (*Schistosoma*) These are extensively found in the tropics where they cause **schistosomiasis** or bilharziasis in up to 200 million people annually.

Life cycle of the blood fluke The adults live in the blood vessels of the abdomen, with the male and female usually lying close together (Fig. 14.4). The eggs bear spines which enable them to work their way out of the blood vessels and into the bladder. They pass out with the urine and hatch into tiny mobile larvae which infect certain species of aquatic snail.

Once established in a snail, they multiply asexually and release a different larval stage called a **cercaria**. Thousands of cercariae may then be released into the water. If these come in contact with a man,

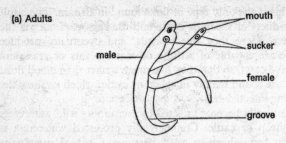

(a) Adults

male

mouth

sucker

female

groove

(b) Life-cycle

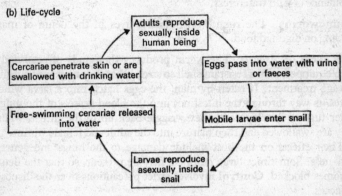

Adults reproduce sexually inside human being

Eggs pass into water with urine or faeces

Mobile larvae enter snail

Larvae reproduce asexually inside snail

Free-swimming cercariae released into water

Cercariae penetrate skin or are swallowed with drinking water

Fig. 14.4 The blood fluke (*Schistosoma*)

they may penetrate the skin and develop into an adult within the blood vessels. Although neither of the free-swimming aquatic stages lives for more than 2 days, the presence of a few infested snails in a pond will result in a large population of cercariae in the water. This creates a high chance of infection for those who drink or enter the water.

The symptoms of schistosomiasis are pain and general weakness caused by damage to the internal organs as the spiny eggs work their way through to the bladder. Pain and bleeding when urinating are common.

Control of Schistosoma involves:

1 hygienic disposal of urine;
2 avoidance of infected waters; infected water should be left to stand for 2 days, after which the cercariae are dead;
3 use of chemicals to kill the snails.

Liver fluke This is the only type of fluke found in Britain. The adult lives in the bile-ducts of sheep or cattle and the eggs pass out via the faeces into water. Here they infect snails, and eventually produce cercariae. These are capable of settling on water plants or grass and forming protective cysts, so that they live much longer than blood fluke cercariae. They are found mainly in sheep or cattle, which swallow the cercariae by eating infected grass.

Humans occasionally become infected by consuming wild watercress growing near sheep or cattle. Commercially produced watercress is always grown far from sheep or cattle pastures. **Control measures** include strict regulation of the growing of commercial watercress and avoidance of wild watercress.

Roundworms The roundworm *Ascaris* lives in the ileum of man, feeding on digested food.

Life cycle of *Ascaris* The worm produces thousands of eggs daily, each equipped with a resistant shell so strong that they may even survive sewage treatment. If eaten by man, the eggs hatch into a larva which bores its way through the intestines into the blood vessels of the lungs. After further development, they wriggle back up to the throat, where they are swallowed and then mature into the adult and female worms.

Their effects on the host include damage to the lungs and general weakness. Sometimes large numbers may be present, so that the ileum becomes blocked. **Control** involves strict precautions over the disposal of faeces.

Pin worms or thread worms (*Enterobius* or *Oxyuris*) are small roundworms living in the colon. They cause little harm except when the female moves out onto the skin around the anus to lay her eggs. Her movements cause great irritation, leading to itching. If scratched, the eggs may become deposited under the finger nails and then swallowed, so that reinfection follows.

The eggs are sticky, so increasing their adhesion to the skin. They are also blown about in dust and can be inhaled or eaten with food. The life cycle of the eggs on a mature adult usually lasts about 5 weeks, so that fresh bouts of irritation occur every 5 weeks.

Pin worms are commonest in children. **Treatment** includes attention to general standards of hygiene in the entire family, e.g. hand washing after use of the toilet and cutting the finger nails. Drugs to kill the worms can be given, but must be applied twice at 3-week intervals to kill both existing adults and the next generation hatching from surviving eggs. The entire household must be treated.

The National Health Service

The National Health Service (NHS) was set up in 1948 to provide free and comprehensive treatment for all. Subsequently, charges for prescriptions, dental treatment, and spectacles were introduced, with exemptions for pregnant women, children under 16, and pensioners.

Organisation of the NHS

The Health Service was reorganized in 1974 as follows:

1 **The Department of Health and Social Services (DHSS)** is the Government Department responsible for the entire service. The Department is allocated funds for the NHS out of general taxation (e.g. income tax) and also from the weekly National Insurance contributions paid by employers and employees.

2 **Fourteen Regional Health Authorities (RHAs)** are responsible for developing long-term plans for the region, allocating money to the different areas and checking on standards within the areas. They are also responsible for the building of hospitals, the ambulance and blood transfusion services and the allocation of specialists within the region.

3 **Ninety-eight Area Health Authorities (AHAs)** in England and Wales provide local services, e.g. hospitals, clinics, family doctor services, etc. Many of the larger AHAs are divided into districts for convenience. The boundaries of AHAs are usually similar to those of elected local authorities (e.g. County or District Councils).

National Services

Services provided by the NHS, for the population generally, include the family doctor and hospital services; treatment by dentists and opticians; and provision of prescribed medicines, at reduced cost.

Services for particular groups include:

Services for pregnant mothers

Ante-natal clinics (also called pre-natal or maternity clinics) provide a full service of checks on the health of the mother and developing baby (*see* p137), and advice on diet, baby care, and family planning. During pregnancy, dental treatment and prescribed medicines are provided free of charge. Vitamin tablets may also be available free.

Financial assistance towards maternity A cash Maternity Grant is paid as a contribution towards the expense of having a baby (e.g.

purchase of clothes, cots, etc.). Working mothers who have paid full National Insurance contributions are also paid a maternity allowance for 17 weeks if they cease work at least 11 weeks before the birth. A woman who has worked for at least 18 months for the same employer is entitled to be re-employed in her previous job 3 months after the baby's birth.

Services provided
Midwives are nurses who specialize in supervising the birth of babies, whether at home or in hospital. They will call for a doctor in the event of difficulties occurring with a birth.

Health visitors are also nurses with a special qualification. They visit homes to advise on care of young children, and sick, old, or handicapped people. They also give advice on how to prevent the spread of diseases such as TB or sexually transmitted infections (VD).

Their main activity centres around very young children. Health visitors are required to visit every family with a new-born child to advise on baby care, look for any signs of abnormal development, and encourage the mother to attend clinics and make use of available vaccinations.

District Nurses are also available if special nursing is required following a difficult birth (or at any other time). Social workers, provided by Local Authorities, will help with social problems and can arrange grants for warm clothing, blankets, and fuel if essential.

Services for young children
Six weeks after birth, the mother and child should attend for a **post-natal examination** by the family doctor, or by a hospital doctor. This check is to ensure that the mother has recovered from the birth and that the baby is developing normally.

Child Health Centres (Child Welfare Clinics or Baby Clinics) Further visits to the Health Centre are desirable so that regular checks can be made on the child's growth in height and weight and development in other ways. Abnormalities detected early on can often be treated before permanent damage occurs (e.g. phenylketonuria, p141). The clinics also provide advice, vitamin drops, and vaccinations (p62).

Services for school-age children
School health services All children are inspected at entry to primary school (age 5) and between the ages of 12 and 14. The inspections include testing of eyesight, hearing, and a dental inspection. In addition, the school nurse will see children from time to time to inspect the hair

for lice, especially the eggs ('nits'), and to check on general hygiene, posture, and development of feet. School doctors also visit schools regularly to examine children referred to them by school nurses, parents, or teachers.

Local Authority services

Elected County and District Councils provide a wide variety of services for the physically and mentally handicapped, and the elderly. These are paid for by grants from central government and from local taxes, known as rates.

Services for the handicapped

Services for the physically handicapped (e.g. the disabled) include help with aids to walking, adaptations to homes, and provision of home helps. Services for the mentally handicapped include provision of special schools and training centres for adults. The training centres provide opportunities to carry out light industrial work in a sheltered environment.

Services for the elderly

These are becoming more important as the number of people aged over 65 increases. The elderly need special attention because the natural process of ageing has the following effects, among others:

1 the bones become brittle, and break easily;
2 the senses deteriorate, leading to deafness, poor vision, and poor balance;
3 the brain ceases to function efficiently, causing loss of memory, and muddled thinking.

As a result, old people may become accident-prone and require supervision to protect them from injury, especially where they live alone.

Services for the elderly include:

1 **Home helps** who assist with the running of a home, e.g. cleaning and preparing meals. They are also available to families of any age, for example where the mother is ill.

2 **Meals services** ('meals-on-wheels') Hot lunches are delivered to homes for a small charge.

3 **Day centres** Old people, and others in special need, are able to attend centres open during the day mainly to provide companionship, light activities, and subsidized meals.

4 Old people's homes provide residential accommodation for those who can no longer live independent lives.

Questions

1 *Name two diseases of man and two human diseases caused by bacteria. Of the four diseases you name, choose two, and give an account of the symptoms, transmission, and the ways in which immunity may be developed for each.* [p164; 2, 2, 6, 6, 9]   (OX)

2 *Diphtheria, smallpox, plague, and tuberculosis used to be widespread diseases in Britain. Today the first three of these are non-existent or extremely rare, while tuberculosis has been greatly reduced in this country. What have been the effective measures, in each case, for this improved situation?* [p164; 14]   (CAM)

3 *With reference to a named parasite of Man, describe* (p169):
   (*i*) *its life cycle* [10];
   (*ii*) *its effect on its host* [2];
   (*iii*) *methods of control* [5];
   (*iv*) *any special features by which it is adapted to a parasitic way of life* [4]

Notes on answer: It is important to choose a parasite which 'fits' the question. Some parasites do not appear to be markedly different from their free-living relatives (e.g. thread worms). It is best to choose a species which is clearly adapted to its way of life, e.g. the tapeworm. The time spent on each part of the answer should be proportional to the marks awarded.

4 (*a*) *Describe the life cycle of the housefly.* [p167; 8]
   (*b*) *Describe how* (*i*) *its structural features and* (*ii*) *its habits, contribute to the spread of disease.* [12]
   (*c*) *How is it possible to reduce the risks from fly-borne diseases in the home?* [5]

5 *What medical and other services are provided for pregnant women up to the birth of the child? Explain the significance of the available provision.* [p173; 20]

Notes on answer: This question calls for a full description, with reasons, of the work of ante-natal clinics, together with the other services mentioned on p173. You should also include the role of midwives and the provision of hospital facilities for the birth itself.

6 *Which government department is responsible for the National Health
Service? How is the service financed? Which of the services are the direct
responsibility of the Regional Health Authority? From the time a child is
born until he or she leaves school as a teenager, which services are available
to maintain health?* [p173; 1, 6, 6, 12]   (ox)

15 Environmental health

Food poisoning and food preservation

Human foodstuffs must be protected from (a) the pathogens which cause food poisoning, and (b) the saprophytes which cause food to decay.

Food poisoning

Certain bacteria found in human food cause the illnesses known as food poisoning, mainly by producing dangerous toxins. The bacteria concerned are commonly found in the soil, and in meat and other foods. They are harmless in small numbers but cause symptoms if allowed to multiply. They include:

1 **Salmonella bacteria**, which cause severe diarrhoea and vomiting. This may occur suddenly, a few hours or days after infection.
2 *Clostridium welchii*, a bacterium which forms spores and is therefore resistant even to boiling. This causes a similar but milder illness than Salmonella.
3 Another bacterium, similar to *Clostridium welchii*, causes the very serious disease known as **botulism**. The botulism bacterium produces a toxin which affects the working of the nervous system.
4 **Staphylococci**, which are bacteria normally found in boils or infected cuts. These may cause diarrhoea and vomiting, if allowed to infect food in large numbers.

Salmonella bacteria are commonly found in the intestines of healthy animals, e.g. sheep, pigs, cows, and poultry. They are also excreted in rat and mouse droppings and can be carried by flies.
Clostridium bacteria are found widely in soil, sewage water, and the intestines. Food is easily contaminated by dust or unwashed hands, and since the spores can survive boiling for 4 hours, even cooked food is

not safe. *Clostridium welchii* is anaerobic, so that it is especially likely to breed deep inside a joint of meat which has been left to cool slowly.

The bacteria which cause **botulism** are also widely distributed, anaerobic, and spore-forming. Fortunately their very poisonous toxin is rapidly destroyed by boiling, and they cannot multiply in acid conditions. Botulism occurs very rarely nowadays; but only because manufacturers are well aware of its dangers.

Staphylococci usually infect food when it is handled by people with unprotected boils, cuts, or sore throats. They multiply rapidly in warm surroundings, but do not breed in acid conditions. Staphylococci especially infect manufactured dishes such as cold meats or pies, and also custard and cream. These foods are likely to be handled after preparation.

Prevention of food poisoning

Temperature control is the first principle of food hygiene. Bacteria can only breed at temperatures between 10 °C and 63 °C, so food should be kept either very hot or very cold. Lukewarm food is extremely dangerous – even after cooking, since both Clostridium spores and the Staphylococcus toxin are unaffected by boiling for short periods. Food must therefore be cooled as quickly as possible after cooking.

Choice of cooking method is also important. Boiling will kill Salmonella but not Clostridium spores, nor will it destroy the Staphylococcus toxin. Roasting and pressure cooking both result in higher cooking temperatures (160–200 °C), which will destroy most microbes and their toxins.

Frozen food must be cooked carefully since freezing merely renders microbes inactive. Large frozen joints or whole birds (e.g. turkeys) may take up to 48 hours to thaw. If cooked while still frozen, the temperature at the centre may not rise above lukewarm, so encouraging rapid multiplication of bacteria. Thawing should take place in a fridge to prevent the temperature of the meat rising too high.

Prevention of contamination is the second principle of food hygiene. Anyone handling food must wash their hands after using the toilet and before handling food. Many healthy people are unknowing carriers of Salmonella. Cuts and boils must be kept covered, and food must be protected against dust, rodents, and insects.

Cross-infection must also be prevented. All raw meat must be regarded as a source of infection, so raw and cooked meat must not come in contact with each other. Utensils and kitchen surfaces used to prepare raw meat must be thoroughly cleaned before they are used for cooked food. Raw meat must not be allowed to drip onto cooked meat in a refrigerator. Sinks and draining boards must be kept clean (*see* p161).

Food spoilage

Food spoilage is mainly caused by saprophytic bacteria and fungi, which feed on human foodstuffs and cause them to decompose. In addition, rodents and insect pests eat stored food and may contaminate it with their droppings.

Foods may also spoil owing to **chemical changes** within them. Following the death of the cells, enzymes may be released which damage the tissues and lead to loss of flavour (e.g. strawberries). In addition, fats may become oxidized on exposure to air, causing them to become rancid (e.g. butter).

Food preservation

Various methods are employed to prevent the spoilage of food. Some involve the killing of all microbes present, e.g. canning; other methods prevent the growth of micro-organisms but do not remove them altogether, e.g. freezing and dehydration.

1 In canning and bottling, food is thoroughly cooked to destroy microbes and placed in cans while still hot. The lid is then sealed on. As the food cools, it contracts, leaving a slight vacuum. The can is made from steel coated with tin to prevent rust and is lacquered internally to prevent acids in the food attacking the metal. Canned foods keep for a year or more, but the flavour and some of the vitamins are lost in the cooking.

2 Dehydration Removal of water from food prevents spoilage by bacteria and enzymes in the food. However, dehydration does not kill microbes so they remain capable of activity when water is added.

The simplest method of dehydration is to dry the food in air, especially in sunny climates. This is used for peas, beans, dates, and fish. **Air-drying** is an ancient and effective method, but much of the flavour of the food is lost as a result.

More modern methods include **vacuum-drying**, where the food is placed in a vacuum. Water is lost rapidly by this means, without any need for heating the food. In **freeze-drying**, the food is frozen rapidly and then placed in a vacuum. After dehydration, the food can be stored at normal temperatures. Both vacuum-drying and freeze-drying preserve the flavour of the food to a greater extent than air-drying.

3 Osmotic preservation If sugar or salt is added to food, water is drawn out of the food by osmosis. This prevents the growth of microbes and may kill them. Foods preserved in this way include jam, bacon, and ham. This method obviously alters the flavour considerably, but the altered taste is acceptable to most people.

4 Smoking Food becomes dehydrated if allowed to dry above a smoky fire. Chemicals in wood smoke may also penetrate the food and inhibit bacterial growth. In addition, an impermeable outer layer forms around the food, preventing the entry of further microbes.

5 Freezing Deep-freezing prevents spoilage because all microbe and enzyme activity ceases at temperatures below $-10\,°C$. Flavour is also retained to a large extent. However, rapid freezing is necessary to prevent the build-up of large ice crystals inside the cells. Since freezing inhibits, but does not kill, microbes, frozen food should be eaten as soon as thawed. It should not be re-frozen later.

6 The use of a refrigerator only has a slight effect on food spoilage. The main compartment in a fridge has a temperature between $0\,°C$ and $10\,°C$, which is sufficient for slow microbial growth. The use of a fridge merely slows down decay, so most foods cannot be kept safely in a refrigerator for more than 2 or 3 days.

7 Pasteurization, discovered by the French scientist **Louis Pasteur,** 1822–98) is a method of killing many types of microbe in wine or milk without boiling it. The liquid is either heated to $63\,°C$ for 30 minutes or to $80\,°C$ for half a minute and then rapidly cooled in both cases. This destroys pathogens in milk and most other types of bacteria, without seriously altering the flavour of the milk. However, pasteurized milk is not sterile and is rapidly spoilt by microbes if it is not kept cool.

8 Sterilization is used to destroy all microbes in food (e.g. sterilized milk). **Ultra-heat treatment (UHT)** is used to produce sterilized milk without serious loss of flavour or vitamins, by heating to $132\,°C$ for 1 minute. It is then transferred to sterile containers under aseptic (microbe-free) conditions. UHT milk will keep fresh for 3 months but its flavour is not identical to pasteurized milk.

9 Food additives Various chemicals may be added to foods to inhibit the growth of microbes. Some of these are 'natural' inhibitors, produced during normal biological processes; others are artificial inhibitors not normally produced by living things. They are discussed below.

Natural inhibitors Since many microbes (especially the dangerous botulism bacterium) cannot multiply in highly acid conditions, the presence of acids helps to preserve foods.

The chief **natural inhibitors** are listed below.

1 **Lactic acid** is formed when certain foods ferment, especially in brine. Herrings fermented in brine can later be canned in an acid sauce without further heat treatment.

2 **Fatty acids** are produced during the fermentation of cheese, and help to prevent further decay.

3 **Acetic acid** (the main component of vinegar) is added to meat and other foods as a preservative. Foods preserved in this way are known as pickles. Acetic acid is formed naturally when certain bacteria ferment wine.

4 **Alcohol** is produced naturally during the fermentation of wine in anaerobic conditions. Its presence inhibits the growth of microbes which spoil the wine.

Artificial inhibitors include sodium benzoate, trichloroacetic acid, and sulphur dioxide. All of these inhibit the growth of microbes without affecting the taste of food.

Pollution

The steady increase in world industry has led to the release of many harmful chemicals into the environment. Land, water, and air may all be affected: in many cases the same pollutant affects all three (e.g. sulphur dioxide).

Air Pollution

The effects of silica, coal dust, asbestos, and other dust particles on the chest, and the harmful effects of smoking, are discussed on p55. Other air pollutants include sulphur dioxide and carbon monoxide.

Sulphur dioxide is produced in the burning of coal and oil. Up to five million tonnes are released into the atmosphere annually in Britain. When mixed with water, it forms a weak acid called sulphurous acid. Over many years this attacks the stonework of buildings, causing them to crumble. The acid rain which results may also lower the pH in the soil and in rivers and lakes, stunting the growth of fish and harming crops. The prevailing winds blow most of the sulphur dioxide produced in northern Europe to Scandinavia, where it falls as acid rain.

To prevent sulphur dioxide pollution, factories often install high chimneys, so that polluted air is carried away at a high level. This helps to prevent local pollution, but increases the chances of pollution in more distant areas. A better solution is to wash the fumes from factories in a strong current of water to dissolve out the sulphur dioxide. Legislation to create smokeless zones is also helpful in reducing air pollution in general.

Carbon monoxide is a highly poisonous gas produced in car exhausts. It combines with the haemoglobin in red cells to produce carboxyhaemoglobin so preventing the formation of oxyhaemoglobin. This causes rapid death due to lack of oxygen. Carbon monoxide concentrations may rise to dangerous levels in cities with high traffic densities in narrow streets surrounded by tall buildings. This occurs in Tokyo, where traffic policemen often use oxygen masks.

Other forms of pollution

DDT is a widely used insecticide with long-lasting effects. It has proved highly effective against mosquitoes and other insects, but is now less favoured since it is known to accumulate in the liver of many animals, esepcially predators, e.g. fish and hawks. By eating other animals containing smaller quantities of DDT, large concentrations build up inside the predator's liver, leading to poisoning or sterility.

Nitrates and phosphates, increasingly used as fertilizers on agricultural land, are liable to be carried into lakes and reservoirs. This leads to pollution of lakes because their presence stimulates an enormous growth of algae (minute plants) in the water. When these die, the decomposition of huge masses of plant remains by bacteria and fungi removes much of the oxygen from the water. This leads to the death of fish and other aquatic animals.

Nitrates may also pollute the water supply, and are often converted to nitrites by bacteria in the water. Nitrites are known to cause stomach upsets in babies.

The solution to nitrate pollution is to persuade farmers to use less fertilizer. At present, more fertilizer is used than can be absorbed by the growing crops.

Lead is also released from car exhausts, and is sometimes present in drinking water in houses with 'soft' water where lead pipes are still in use. Lead damages the nervous system, especially in children, and leads to lower intelligence and restless behaviour. It may affect children who live near main roads.

At present, lead is deliberately added to petrol to improve the efficiency of car engines. This practice may have to be banned in future. In homes with lead pipes, the cold water tap should be run every morning to draw off water which has been standing overnight in the pipes. Replacement of pipes is highly desirable.

Metal wastes, e.g. copper, mercury, zinc, and tin, may pollute both land and water. These metals are found in the slag produced from mines or from refining operations, and in the waste water from many forms of industry. Metal pollution is highly poisonous to life: for example, slag heaps in the Swansea area will barely support plant life. Many hundreds of Japanese have been seriously affected by eating fish containing mercury, which damages the nervous system. The solution is to insist on extraction of metals from factory wastes before their release into rivers.

Atomic radiation, produced from atomic explosions and accidental discharge from nuclear power stations, causes radiation sickness. This may produce nausea, sterility, mutations in children, and various malformations. Exposure to high levels of dose is fatal.

One long-term effect of radiation is the release of radioactive material into the human food cycle. For example, radioactive strontium released during nuclear explosions is washed out of the atmosphere and into the soil by rain. Cattle may then absorb it from grass, and children from their milk. The radioactive strontium then accumulates in the skeleton, leading to malformations in the bones.

Complete cessation of all nuclear tests will help to minimize the level of radioactive strontium. Similarly, unnecessary exposure to X-rays, a form of radiation, should be avoided.

Water purification

The purpose of water purification, is to produce potable water, i.e. water fit for drinking. This should be free from:

1 debris, soil particles, etc.;
2 microbes, especially pathogens;
3 industrial wastes, i.e. chemicals of various kinds;
4 undesirable tastes or odours.

The **sources** of drinking water include surface water from springs, rivers, lakes, reservoirs, and shallow wells; and deep water obtained by sinking deep wells into underground water supplies. Water from some deep wells may be pure enough to drink, but water from all other sources must be purified.

Stages in water purification

1 Sedimentation Water is obtained from the purest available source, and pumped in through a **screen** to prevent entry of large objects, including fish. It is then allowed to stand in a settlement tank, to allow large particles to settle out. The water is later pumped from the surface of the settling tank to a filter.

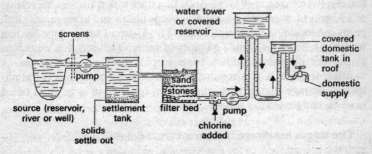

Fig. 15.1 Water purification

2 Filtration takes place either through a slow sand filter or through a pressure filter. **Slow sand filters** have a large surface area and contain layers of sand resting on layers of grit and larger stones. Inside the sand, a natural gelatinous film of slime forms from the secretions of millions of microbes. Organic material in the water is consumed by the microbes as it passes through the slime. Pathogens are also eaten by the millions of protozoa found in the slime.

In a **pressure filter,** water is forced through a smaller, but similar, filter in which a layer of alum is used to form an artificial colloidal jelly. This provides more rapid filtration but is more expensive to use.

3 Chlorination removes any microbes which may have passed through the filter. **Fluoride** salts are also added in some areas, to help prevent tooth decay (p81).

Potable water must be **stored** where there is no risk of contamination. It is usually pumped to high-level storage tanks or to covered reservoirs on hills. It then descends by gravity into the main water supply below. Similarly, domestic water is stored under the roof in covered tanks, which must be proof against the entry of rodents, insects, and birds.

Disposal of human wastes

Treatment of sewage

Human urine and faeces are together known as sewage. Sewage must be carefully treated after collection to remove pathogens.

Hygienic treatment of sewage begins with its collection in **flush toilets**, (water closets or lavatories). In a flush toilet, the sudden release of 15 litres of water from a cistern sweeps faeces and urine rapidly into pipes connected to the main drains. The **U-bend** in the pipe leading from the toilet traps a small quantity of water. This acts as a water seal to prevent obnoxious odours from seeping back into the house. A **ventilator pipe** is also required to release these gases into the atmosphere. Sewage from main drains is then collected in large pipes called **sewers**, and passed to a sewage works.

The stages in sewage treatment are as follows (Fig. 15·2):

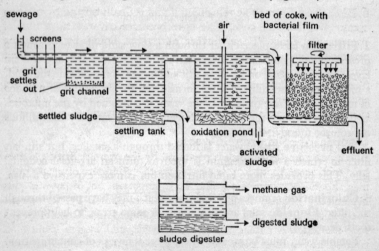

Fig. 15.2 Sewage treatment

1 Screening The incoming liquid is passed through a screen to remove solid objects, e.g. bottles, cans, or babies' nappies.

2 Grit settlement The liquid is passed slowly along a narrow open channel to allow heavy objects, e.g. grit, to settle out.

3 Settling tank In these large tanks, the liquid is allowed to stand long enough for suspended solids to settle out and form settled sludge.

4 Sludge digester The settled sludge is pumped into sludge digestion tanks. These huge vats hold hundreds of gallons of semi-liquid sludge. They are not aerated, so encouraging the growth of anaerobic bacteria which digest the organic material, producing methane gas. This is used to drive pumps in the sewage works and may also be sold. The digested sludge is then dried out and sold as fertilizer.

5 Oxidation ponds (aeration tanks) In a modern sewage works, the liquid is next run into large ponds where the water is constantly stirred and agitated. This encourages the growth of aerobic bacteria which oxidize much of the organic material to carbon dioxide. Protozoa also flourish in these conditions, consuming numerous bacteria, including pathogens. Some organic material also settles out at this stage as activated sludge. This is dried out and sold as fertilizer.

6 Biological filter The remaining liquid is finally spread by rotating sprays over beds of coke. The coke becomes covered by a film of bacteria and fungi. These further purify the liquid by consuming organic material. Protozoa, roundworms, and insect larvae in the filter consume bacteria, so preventing clogging. The perforated sides of the filter tank, and the gaps between the pieces of coke allow entry of air, encouraging the growth of aerobic bacteria. The effluent is now sufficiently pure to enable it to be passed directly into rivers.

Disposal of solid organic waste

About one tonne of rubbish is produced by the average British household each year. About 12% is organic waste – a mixture of potato peelings, bones, left-over food, etc. It is important to dispose of this hygienically because organic wastes are decomposed by saprophytic bacteria and fungi, giving rise to undesirable odours. Organic wastes may also contain pathogens. Rodents, birds, and insects feeding on rubbish may subsequently spread these pathogens to man. Rainwater falling on exposed rubbish may also spread pathogens into rivers and lakes.

Dustbins

The first stage in hygienic disposal is to use properly designed bins. These should possess close-fitting lids to exclude rodents and insects. The lids should be clipped on to prevent dislodging by cats or dogs. The interior must be made from galvanized steel or plastic to provide smooth, easily cleaned surfaces. These should be lined and the lining disposed of with the contents. Bins should also be washed and disinfected regularly.

Refuse collection vehicles should also be fitted with close-fitting covers to prevent accessibility to wind, rain, and flies. Many possess mechanical rams to compress refuse towards the front of the truck.

Refuse tips

Most rubbish in Britain is tipped into hollowed out areas in waste land. It is then compressed into a small volume by bulldozers until it is about 2 metres deep, and covered with up to half a metre of tightly packed soil to prevent access by rodents, insects, and birds. Burial in the ground speeds decomposition by soil microbes.

In addition, rat and insect numbers around tips must be kept down, and high chicken-wire fences are necessary to catch any rubbish blown by the wind. After a hollow has been filled in, it is possible to grass the tip and so reclaim the land for recreation or agriculture.

Pulverization is a method used to break up rubbish into small pieces, by using a machine called a pulverizer. Pulverized refuse occupies less space and can be compressed more easily than untreated rubbish. This makes it less attractive to rats, flies, and gulls. To assist in its decomposition, water or sewage sludge is added.

Recycling of solid organic waste

Organic waste can be converted into compost. After removal of all inorganic material (e.g. cans, plastic, etc.), the wastes are left to decompose in a rotating drum for 5 days. Controlled amounts of sewage sludge are added, to provide a source of saprophytic bacteria, and the drum is aerated to encourage aerobic respiration.

Few pathogens can survive this process of decomposition, and the resulting material has little food value for rats, insects, and gulls. However, it is an excellent fertilizer which improves the moisture-holding capacity of poor soils and provides minerals for plant growth.

Role of Environmental Health Officers

Several laws have been passed to compel householders, local authorities, factory managers, and especially caterers, to observe the basic principles of hygiene. To ensure that the laws are obeyed, and to give advice, district councils employ Environmental Health Officers (EHOs), formerly called public health inspectors. Their duties include:

1 Housing EHOs inspect houses, especially hotels, etc., to check on overcrowding, the presence of vermin, and the disinfecting of clothes from persons with infectious diseases.

2 Water supply and waste disposal EHOs sample water supplies to ensure that the water is potable; that sewage is efficiently dealt with; and that rubbish tips are properly constructed and free from vermin.

3 Pollution EHOs check on noise pollution, and on air pollution from factory chimneys.

4 Food hygiene EHOs visit food shops, cafés, caterers' premises, and food factories to check that the premises are well lit, clean, and ventilated; that toilets and washbasins are provided for staff; and that equipment used in the preparation of food is kept clean. They also check on the personal hygiene of food handlers and on the storage of perishable food at correct temperatures.

If EHOs discover a breach of the laws, they issue advice and warnings. In serious cases, they may recommend prosecution.

Housing

Houses are essentially a form of shelter. They are constructed in order to protect people from rain, wind, extremes of heat and cold, and unwanted animals. Their design varies according to the climate and the availability of local building materials.

Waterproofing
To keep out rain, sloping roofs are covered with overlapping tiles or slates, and flat roofs with roofing felt. Sloping roofs are essential in climates with a heavy snowfall, since an accumulation of snow might cause the roof to collapse.

To prevent rain from penetrating the sides of the house, the walls are made of two layers, separated by a cavity. The cavity prevents moisture from crossing to the inner layer.

Prevention of rising damp (Fig. 15·3)
Most building materials (e.g. bricks or timber) absorb water by capillary action. Moisture from the ground tends to rise inside the walls, so causing permanent dampness, spoiling decorations and encouraging the growth of fungi which rot the timbers in the house.

To prevent rising damp, a **damp-proof course** is installed low down in the walls. This consists of a layer of impermeable material, e.g. slate or bituminous felt, which prevents moisture from passing up the walls. It is important to prevent soil from piling up against the walls above the level of the damp-proof course. Damp soil may form a bridge from the ground, across the damp-proof course.

To prevent ground moisture affecting the floors, the floor boards are raised above the ground on joists set into the walls. To keep the floor

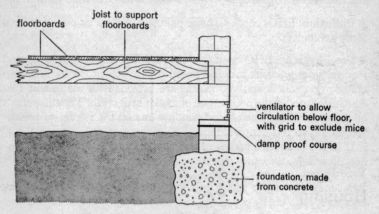

Fig. 15.3 Foundations of a house

dry, air is allowed to circulate in the cavity below the floor boards, through **ventilation bricks**. It is important to ensure that the ventilation bricks do not become blocked with soil.

Insulation

In most modern houses, the walls consist of two layers of bricks with an air-filled **cavity** between them. This provides a high degree of insulation because still air is a poor conductor of heat. The insulation can be further improved by injecting the cavity with foam.

Since warm air rises, much heat is lost through the roof. The floor of any **loft space** below the roof should therefore be thoroughly insulated with glass fibre wadding, or granulated polystyrene (Fig. 15·4)

Windows can be insulated with **double glazing**. This is based on the same principle as cavity walls: by adding an extra layer of glass, still air is trapped between the panes.

In tropical climates, it may be more important to lose heat. If so, the house may be built of lighter materials with an open roof space and roof ventilators. The roof should be made of shiny materials to reflect the sun.

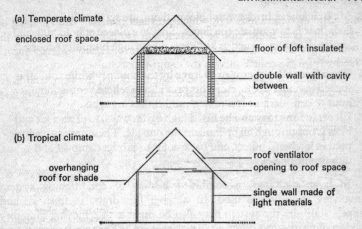

Fig. 15.4 Design of houses in temperate and tropical climates insulation and ventilation

Ventilation

Regular changes of air inside buildings are essential. Air becomes stale because it is altered by people in these ways:

1 The oxygen content falls and the carbon dioxide content rises, owing to respiration.
2 The humidity increases as a result of breathing and perspiration.
3 The temperature rises (three people give off as much heat as a 1 kilowatt electric fire).
4 The concentration of micro-organisms in the air rises, so increasing the risk of infection.

The most serious of these changes is the rise in humidity. In a damp atmosphere, sweat will not evaporate, so that the body's normal cooling mechanism fails to work. This leads to feelings of irritability and lack of concentration.

Methods of ventilation Since warm air rises, most methods of ventilation include openings high up in rooms to allow for the escape of warm stale air. Other openings at a low level allow cold, fresh air to enter. The commonest methods of ventilation are:

1 **Windows** All windows should have at least one section which can be opened. The small upper sections of windows (fanlights) can be left open to allow loss of stale air without creating an irritating draught.

2 **Air bricks,** i.e. bricks with holes in them, are set in walls below
   floor boards to ventilate the floors. They are placed at both high and
   low levels in rooms such as kitchens, where high humidity occurs
   regularly.
3 **Open fires** allow warm air to leave by the chimney while cold air is
   sucked in under doors, etc. This provides excellent ventilation but
   leads to considerable loss of heat, and causes draughts.
4 **Electric fans** may also be fitted high up on walls, to extract air and
   blow it out through an opening to the outside. These are especially
   used in kitchens, toilets, and bathrooms to extract damp air and
   unpleasant smells.

In large buildings, a combined heating/air conditioning system
involving large fans is used. In summer this draws in fresh air and
expels stale air: the system may also be capable of removing humidity
and cooling the incoming air. In winter, incoming air is heated to a
comfortable temperature.

Heating

In cold climates, heating must be provided during the winter. Generally,
living rooms are kept at 15 to 20 °C, and bedrooms at between 10 and
15 °C. The various types of heating are shown in Table 15·1.

Central heating is increasingly used to heat an entire house, and is
essential for large buildings. Typically, it involves a gas, coal, or oil-
fired boiler which heats water to about 82 °C. This is then pumped
around the house to radiators in each room. Alternatively, the boiler is
used to provide warm air which is blown around the house, entering
each room through under-floor ducts.

Central heating is much the most economical and efficient way to
heat a complete house or building. It provides an even temperature
which is especially desirable for the very old, the sick, or the young.
However, it is difficult to control since the temperature in individual
rooms cannot easily be adjusted. In addition, ventilation must be
provided separately.

Convection, conduction, and radiation   All of the methods of
heating mentioned provide heat by convection, i.e. they create warm
currents of air which rise, and are replaced by colder air. This creates a
circulation of air in the room. Coal, gas, and electric fires also provide
radiant heat. In the case of electric fires, radiation is the main source of
heating. None of these methods provides much heat by conduction
because air is a poor conductor of heat.

Table 15.1 Comparison of methods of heating

Method	Advantages	Disadvantages
1 Coal fire (solid fuel)	Creates a warm and cheerful atmosphere. Ventilation must be provided.	Inefficient, as most of the heat is lost up the chimney. Considerable work is required in fetching coal and removing ash. Smokeless fuels are necessary to avoid air pollution. Ash leads to dirt inside the home.
2 Gas fires	Cleaner than coal, easy to light and control. Ventilation must be provided.	Any leak may cause an explosion.
3 Electric fires	Clean, easy to control, but expensive to use. Portable.	No ventilation required for their operation, so this must be provided separately.
4 Paraffin heaters	Much favoured in the past because of cheapness of fuel. Portable	Serious fire hazard if knocked over. Give out an unpleasant smell if not kept clean.

Lighting

Natural lighting should be used wherever possible, so that in temperate climates windows should occupy at least 10% of the wall surface. Access to natural light also improves morale.

The type of lighting required depends on the purpose for which it is needed. Strong fluorescent lights are useful in workrooms, e.g. kitchens, and in factories, since they cast no shadows. Diffuse light is preferred for living rooms since it is more attractive to the eye. Good reading lights must also be provided, especially for old people. Stairs and corridors must be well lit to prevent accidents.

Town planning

In designing new towns or housing estates, the following principles must be borne in mind:

1 **Marshy ground** should not be chosen for building land. If it is used, extensive drainage will be necessary to prevent the foundations from sinking, and the house from becoming damp.

2 **Factories** should be sited so that the prevailing wind does not blow polluted air over houses.

3 **Playing fields,** parks, and children's play areas must be provided for recreation.

4 **Ample open space** should be allowed around the estate to allow penetration of sunlight, so encouraging the manufacture of vitamin D in the skin.

5 **Homes should be built away from main roads** to avoid fumes, noise, and lead poisoning from car exhausts.

6 **Multi-storey tower blocks** should be used as little as possible. Although these offer great savings in space, they are not suitable for families owing to the feelings of isolation which result and the difficulty in sending young children out to play.

7 **Sewage works** should be situated down-stream from water purification plants, so that there is no risk of sewage effluent being taken into the purification plant.

Questions

1 *Describe concisely the precautions that should be taken in the handling, preparation and storage of food, to prevent outbreaks of 'food poisoning' and the contamination and deterioration of food.* [p179; 15] (CAM)

2 (a) *What types of agent cause food to decay?* [p178; 2]

  (b) *List four methods of food preservation. For each method explain why the food is prevented from going bad, and state the advantages and disadvantages of each method* [p180; 16]

  (c) *If decaying food is thoughly cooked before eating, does this render it completely safe? Explain your answer.* [p179; 2] (JMB)

3 *In what ways are foods likely to deteriorate other than by the action of microbes?* [p180; 4]

4 *Write as essay on pollution of air, water, and land. Explain the harmful effects of pollution and suggest how it may be controlled.* [p182; 20]

Notes on the answer: This answer requires careful planning It is best to concentrate on a few examples and deal with each in detail, rather than quoting numerous pollutants and saying little about each.

Plan: Sulphur dioxide: air/water/soil: effects on stonework/crops/fish; control methods.

  Nitrates and phosphates: water: effects on fish and babies; control methods.

  Atomic radiation: air, milk; effects on body; control methods.

If you have a choice, try to avoid answering general questions like this. It is easy to write a long, vague answer which will win very few marks.

5 *How is water purified to make it fit for drinking? Give a reason for each stage in the process.* [p184; 15]

6 *Give two reasons why a water trap or U-bend is necessary in toilets and bathrooms.* [p186; 2]

7 *Draw and label two diagrams to show:*
   (a) *a section through a filter bed at a sewage works.* [p186; 6]
   (b) *a section through a purifying reservoir at a water works.* [p185; 6]
   Indicate clearly how the liquid is introduced into each tank, its passage through the tank, and its exit. Explain in each case where the liquid is finally taken and the biological principle involved in its purification. [4] (WJ)

8 *Describe the methods currently used for the disposal of dry refuse in towns.* [p187; 10]

9 *Briefly describe the work of environmental health officers.* [p188; 10]

10 *Describe how a modern house (p189) in Britain should be: (a) ventilated [5], (b) insulated [5], (c) rendered waterproof.* [10]

Index

Index